BRITAIN, SPAIN AND GIBRALTAR 1945–90

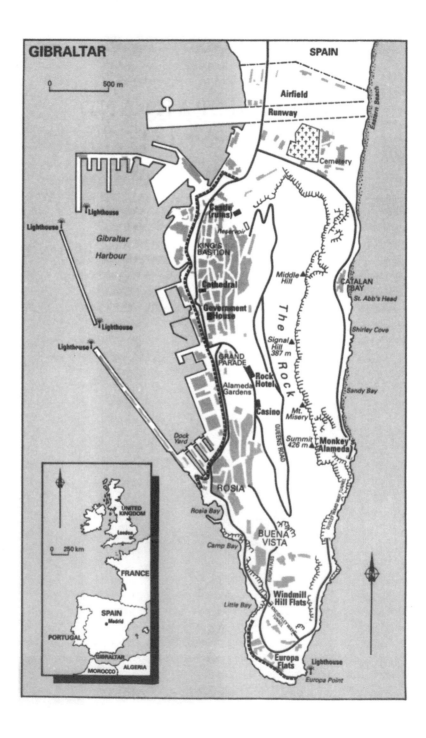

BRITAIN, SPAIN AND GIBRALTAR 1945–90

The Eternal Triangle

D.S. Morris and R.H. Haigh

Routledge
Taylor & Francis Group

LONDON AND NEW YORK

First published 1992
by Routledge
2 Park Square, Milton Park, Abingdon, Oxfordshire OX14 4RN

Simultaneously published in the USA and Canada
by Routledge
711 Third Avenue, New York, NY 10017

First issued in paperback 2014

Routledge is an imprint of the Taylor & Francis Group, an informa business

© 1992 D.S. Morris and R.H. Haigh

Typeset by Selectmove Ltd.

British Library Cataloguing in Publication Data
Morris, D. S. (Dennis S), *1945–*
Britain, Spain and Gibraltar 1945–1990 :
the eternal triangle.
I. Title   II. Haigh, R. H. (Robert Henry), *1935–*
946.89
0–415–07145–3

Library of Congress Cataloging in Publication Data
Morris, D. S.
Britain, Spain, and Gibraltar, 1945–1990 : the eternal triangle /
D.S. Morris and R.H. Haigh:
p.   cm.
Includes bibliographic references and index.
ISBN 0–415–07145–3

1. Gibraltar—International status.   2. Great Britain—Foreign
relations—Spain.   3. Spain—Foreign relations—Great Britain.
I. Haigh, R. H.   II. Title.
JX4084.G5M67   1992
341.2'9'094689—dc20   91–20990
                                       CIP

ISBN 13: 978-0-415-07145-1 (hbk)
ISBN 13: 978-0-415-75587-0 (pbk)

For Quango,
A loyal and learned friend who went in search of
greater knowledge on 28 November 1990

CONTENTS

FOREWORD

This book provides a well researched and comprehensive account of Gibraltar's post-war political history and development. It should not only be a valuable source of reference but also a testimony to the manner in which the Gibraltarians have worked their way through to democratic self-government against difficult odds.

The book should appeal to observers, commentators and students of Gibraltar and its people. It provides detailed evidence in support of the argument that through years of struggle the Gibraltarians have earned the right to self-determination and to be the sole arbiters of their own destiny.

Rt Hon. J.J. Bossano
Chief Minister
Government of Gibraltar

ACKNOWLEDGEMENTS

Authors always incur debts and we are no exception to this seemingly universal rule. The only repayment we can make for the help which we have received is to offer our sincere thanks to the following without whom this work would never have come to fruition.

Foremost we must acknowledge our indebtedness to the British Council which supported our request for an award under the Acciones Integradas scheme. Without this financial assistance our task would have been rendered incalculably harder if not impossible. We owe our appreciation to the librarians of the Eric Mensforth Library of Sheffield City Polytechnic for the patience they showed when faced by our seemingly endless requests for material and for the diligence which they exhibited in meeting every demand with praiseworthy despatch.

To the politicians, businessmen and trade unionists of Gibraltar we owe a debt for which mere words seem a churlish recompense, but words of thanks is all we can offer in return for the tolerance with which they greeted our questions, for the detail, clarity and honesty of their answers and for giving so unhesitatingly and courteously of their time. In particular, we would wish to record our sincere thanks to: Mr Walter Prior, Press Officer Gibraltar Government, for arranging a schedule of interviews which met our every need; The Rt. Hon. Joe Bossano, Chief Minister of Gibraltar, for talking to us so openly and at such length at a time when he was confronted by a work load which would have consumed every minute of the working day for a less energetic and committed politician; The Rt. Hon. Joe Pilcher, Minister for GSL and Tourism and Deputy Chief Minister, for the insights with which he provided us into the economy of the Rock; The Rt. Hon. Adolfo Canepa and the Rt. Hon. Peter Montegriffo, Leader and then Deputy Leader respectively of the Opposition for the understanding they gave us of the trends in Gibraltarian politics; Mr Joe Gomez and Mr Jose Netto, of the Gibraltar Branch of the Transport and General Workers Union and Mr Edward Navas, of the IPCS, for affording us an appreciation of trade union issues on the Rock which was unsurpassed for its depth and clarity; and to Mr Sol Seruya and Mr Lewis Andlaw, then respectively President and Vice President of the Gibraltar Chamber of Commerce, for explaining so cogently the difficulties encountered by the community's businessmen during the period of partial and total frontier closure. To all of the above we owe thanks which we feel ourselves unable to adequately repay.

We must also express our appreciation to Tristan Palmer and Jennifer Binnie of Routledge for their encouragement, support and professionalism, all of which made the production of a final manuscript a much easier and more pleasurable task than would otherwise have been the case.

Finally, we thank our families for once again tolerating our preoccupation with the task in hand. We must have come close to boring them beyond belief with our constant talk of Gibraltar and the unsolicited bulletins which we gave them of our progress. We have been assured that although we have put them through similar trials on many previous occasions we have not become any easier to live with when engaged in research. We thank them for their tolerance and good humour in return for the neglect which we have bestowed upon them.

The real heroes and heroines of Gibraltar must remain the Gibraltarians themselves for their resolution in the face of prolonged adversity and for the energy which they are showing in rebuilding their community on a foundation worthy of the future they are seeking. To them we dedicate this work in the hope that it will serve as some small tribute to their courage, fortitude and zeal.

D.S. Morris, R.H. Haigh

1

THE AWAKENING PROBLEM

Even before the Second World War reached its conclusion there was an acceptance in Britain that the pre-war relationship that had existed between the mother country and her empire could not be re-established on its old basis and according to its previous tenets. New links would have to be forged to accommodate the new circumstances which global conflict, on a previously unheralded scale, had occasioned. The magnitude of change – political, social and economic – resulting from the second major conflagration of the twentieth century necessitated the granting of political advancement to territories which still lay under British colonial administration. Gibraltar was not to prove an exception to this general trend and on 29 December 1944 constitutional changes affecting the Colony were announced.[1]

For Gibraltar, constitutional changes could be viewed as reward for the part played by the Colony throughout the wartime years. A bastion guarding the Atlantic approaches to the Mediterranean Sea, the Colony had been a key factor in both the North African and Italian campaigns. Increased autonomy constituted not only recognition of that contribution but also served as recompense to the citizens of Gibraltar for the suffering they had endured between 1939 and 1945. The fortress nature of Gibraltar, allied to its strategic position, had made necessary the introduction of a compulsory evacuation programme for its non-service personnel. The programme began in the late spring of 1940 and involved the evacuation of some 15,000 civilians, mainly women and children, who were despatched to such disparate destinations as the United Kingdom, Madeira, Jamaica and Tangier.[2]

By April 1944 the majority of the evacuees had returned home but a sizeable minority remained overseas and a demonstration calling for a hastening of the repatriation process took place. On 25 February 1945, Mrs Horsburgh, Parliamentary Secretary, Ministry of Health, told the House of Commons that 5,800 of the Gibraltarian evacuees were still in the United Kingdom and that their departure for home was being delayed by a lack of shipping and a shortage of housing accommodation in Gibraltar. This minority group had initially been evacuated to the London area but had been relocated to reception camps in Northern Ireland when the capital came under attack from V-bombs.[3] Indeed, as the war ended, the delays to the repatriation programme seemed to suggest that the relationship between Gibraltar and Britain was more acrimonious than the relationship between Gibraltar and Spain.

Although Britain had emerged victorious from the Second World War she had been much weakened, industrially, commercially and economically by her exertions, and her former political eminence as a world power of the first order was reduced still further by the growing international ascendancy of two of her wartime allies: the United States and the Soviet Union. Demands for increasing political independence from her colonial possessions seemed certain to ensure that Britain would have to seek a new role in world affairs; a role which would, from necessity and expediency, lack the grandeur of earlier decades.[4]

Nevertheless, Phillip Dennis expresses the view that:

> At the end of World War II the Rock seemed more British than ever before ... and the idea that Spain might even consider claiming sovereignty over Gibraltar was far removed from anybody's thoughts in Britain.[5]

Such a conclusion, however, depends for its validity more on an assessment of Spanish weakness than from an assessment of British self-confidence; a weakness originating in the turmoil of civil war[6] and compounded by Spain's relative international isolation in the immediate post-1945 years. Spain's exclusion from the community of nations was itself a consequence of her non-belligerent but pro-Axis orientation during the Second World War and her form of governance: a military dictatorship much out of keeping with the spirit of increased democratisation which was high on the agenda of post-war international political life.[7]

Medhurst offers a pertinent summary of Spain's situation with the return of peace:

During the 1940's the central issue in Spanish politics was the Regime's survival. Its defeated enemies were in no position to fight back but they hoped that a combination of international pressures and internal divisions would provoke a collapse. Abroad the regime's reputation was mortgaged because of its links with the Axis powers. . . . After 1945, therefore, Spain was subjected to diplomatic isolation and economic blockade.[8]

Paradoxically, those international pressures which might have served to engender political change in Spain enabled Franco to portray himself as a symbol of Spanish national interest in the face of external hostility and thus to consolidate his position, and the regime which he had founded, at a time when it appeared to be most vulnerable to internal divisions and factionalism.[9]

The growing intensity of Cold War antagonisms between East and West in the late 1940s and into the 1950s served to lessen the hostility felt towards Spain by the democratic members of the international community. As Medhurst notes:

Such developments signalled an end to Spain's isolation. At home they killed any doubts about the regime's durability and respectability. In the 1950s few continued to expect a quick collapse.[10]

Spain's first readily discernible step away from international isolation and back into the community of nations came with the signing by the Spanish Foreign Minister, Señor Martin Artajo, and the US Ambassador, Mr James Dunn, in Madrid on 27 September 1953, of three bilateral agreements between the United States and Spain which provided for the construction and use of military bases in Spain by the USA, US economic assistance to Spain, and US military supplies to Spain. Washington stressed that these agreements were executive agreements and not a treaty of alliance requiring Senate approval.[11]

No official details were given as to the locations of the US bases on Spanish soil but press reports in America contained speculation that initially airfields near Madrid, Barcelona and Seville and naval bases at Coruna, Cadiz and Cartagena were likely to be utilised. Work on the bases was to begin immediately with monies drawn from special funds made available by the US Defense Department for the construction of overseas installations.

Although the three bilateral agreements between the US and Spain did not require approval from the American Senate they, nevertheless, received warm support from several leading Senators. However, less enthusiasm was afforded the agreements by certain sections of the American press. In particular, the *New York Times* exhibited some trepidation when it commented:

> We are now faced with the necessity of swallowing a bitter pill – the military agreement with Franco Spain. Let us all hope that the medicine will do more good than harm.[12]

Not surprisingly, Franco expressed no such reservations and in a message to the Cortes, on 30 September, he hailed the agreements as 'the most important achievement of our contemporary foreign policy'. Arguing that Spain, from the time of the civil war, had been the only European nation to consistently warn of the danger of communism, Franco conceded that during and after the Second World War, Spain had appeared to be alone in pointing to the prospect of communist expansion and that this had led to her isolation from the mainstream of European political life because 'of the blindness with which our aims were received'. Confronted by the 'persistent hostility and incomprehension of the European Powers' Spain had orientated herself towards the more perceptive and receptive United States. Pointing to the fact that Soviet communism would not cease its expansion at the Spanish frontier, Franco contended that the agreements signed with the US would produce the external collaboration necessary for Spain to prepare her defences, without implying that she was dependent on other nations to defend her, whilst at the same time consolidating the strategic unity of the Iberian Peninsula which had been initiated with the signing of the 1939 Treaty of Friendship and Non-aggression with Portugal.[13]

The agreement was described in Paris by a spokesman for the Quai d'Orsay as a 'purely Spanish–American affair' before adding that the French Government had been kept informed of the negotiations by the United States. In London, there was no comment from official sources but *The Times* pointed to some of the wider implications of the agreement:

> The European members of NATO are bound to be affected by this direct and close military agreement between the leading member of the alliance and a country of great strategic concern to all members alike. All of Spain's neighbours are within the Treaty; the seas around her are covered by NATO commands ... The objections likely to be heard will be mainly political

... The ground for this agreement is military necessity ... it is not the first time that a military necessity has been regretted by many on other grounds.[14]

In Moscow there was condemnation of the agreement which was described as 'an open military alliance between the USA and Franco Spain' and as 'an attempt on the part of the USA to bind Franco Spain de facto, if not de jure, to the war bloc in Europe'.

A direct consequence of this eradication of internal and external pressure upon the Franco regime was a growth in the regime's self-confidence and self-assertiveness and one aspect of this new found vigour was seen in the renewal of Spanish claims to Gibraltar.

Following her coronation in June 1953, Queen Elizabeth, accompanied by the Duke of Edinburgh, embarked on a six-month Commonwealth tour in the autumn which was to culminate in a visit to Gibraltar on 10–11 May 1954. On 17 January the British Foreign Office revealed that five days earlier the Spanish Ambassador in London, the Duke de Primo de Rivera, had called on the Foreign Secretary, Mr Anthony Eden, and requested that the Queen's projected visit to Gibraltar be cancelled. The Ambassador had been informed that the British Government could not countenance representations from a foreign power regarding visits made by Her Majesty to any of her territories.

Two days after the disclosure in London of the meeting between the Spanish Ambassador and the British Foreign Secretary, the Spanish Foreign Ministry, in Madrid, issued a statement about that meeting in which it was pointed out that the Duke de Primo de Rivera had indicated the 'resentment' felt by the Spanish people that 'the fortress of Gibraltar' had been included in the itinerary of the Queen's Commonwealth tour and had intimated that such an event would be 'imprudent', that it might have an adverse effect on Anglo-Spanish relations, and that it would inevitably call forth a 'national protest' from the Spanish people given that 'Gibraltar is Spanish territory to which the Spanish people do not renounce their claim'.

Such a statement proved to have all of the weight of a self-fulfilling prophecy as widespread anti-British demonstrations by Spanish university students took place on 22 and 25 January in several cities including Barcelona, Cordoba, Granada, Madrid and Seville.

In Barcelona several hundred students staged anti-British demonstrations during the course of which windows were smashed at the British Consulate and British Institute. In Cordoba, students paraded through the streets demanding that Gibraltar be returned to Spain while in Granada windows were broken at the British Vice Consulate and the Spanish flag was raised over the building. The most serious disturbances were reserved for Madrid where, on 22 January several thousand students carrying Spanish and Falangist flags demonstrated outside the British Embassy and smashed windows before being dispersed by police. A second, and more violent demonstration, took place three days later in Madrid when a crowd, estimated at 30,000, repeatedly charged and stoned a strong police presence protecting the British Embassy. The police retaliated by firing blank cartridges into the air and by launching a series of baton charges which left 30 demonstrators and 18 policemen injured and in need of hospital treatment. A number of arrests were also made. During the demonstrations anti-British slogans were chanted, demands made for the return of Gibraltar to Spain, Falangist Party songs sung and banners advising, 'Cuidado, Reina Isabel' were prominently displayed. In Seville, a crowd of some 3000 students threw oranges at the British Consulate and cried, 'Franco, Franco. We want Gibraltar'.

Sir John Balfour, the British Ambassador in Madrid, lodged two strong protests with the Spanish Government over the demonstrations and damage done to British property, without receiving a reply. However, the Spanish Government did issue a statement on 28 January saying that the Madrid demonstrations had been initially inspired by 'high patriotic motives' but had led to 'distressing incidents which are deplored by all'. The statement went on to say that inquiries were being undertaken to establish the facts and where responsibility for the riots lay but that it was already clear that 'extraneous elements . . . had mixed with the students with the aim of spoiling their noble intentions'. On the same day the British Admiralty announced in London that the Home Fleet had cancelled visits to have been made by some naval units, in the spring, to Spanish and Spanish Moroccan ports. Plans for the Queen's visit to Gibraltar remained unchanged.

However, extensive security precautions were taken in Gibraltar for the royal visit in view of the Spanish request that it be cancelled and in the wake of the anti-British demonstrations. On 1 May the Spanish Consulate in the Colony was closed. The reason for this decision was given as being that the volume of business transacted no longer justified its retention, but other reasons were also suggested, among these being that in the circumstances it would be impossible for the Spanish Consul to attend official functions or fly the Spanish flag over the Consulate during the Queen's visit. On 7 May all of the 12,000 Spanish workers who daily crossed the frontier to work in Gibraltar were subjected to a thorough documentation check and a personal search, and during the two days of the Queen's visit no exit permits were issued by the Spanish authorities except at the request of the Gibraltarian authorities. Those who were allowed to enter the Colony from Spain, chiefly hotel workers and those engaged in essential public service, were not permitted to return to Spain until after the royal party had left Gibraltar.

By the mid-1950s, therefore, Gibraltar had clearly become an issue on the Anglo-Spanish agenda but it was an issue which neither party to the relationship seemed over-anxious to pursue. Britain was clearly content to maintain the status quo over Gibraltar whilst Spain confronted economic problems which increasingly came to the forefront of her domestic political debate. As Medhurst notes:

> international circumstances and the regime's revivalist temper impelled (Spain) to adopt a policy of autarchy . . . requiring massive state intervention. The economy's basic structure, however, remained unchanged, and when . . . there was a measure of economic recovery, its basic weaknesses became apparent.[15]

Autarchy, the subject of criticism in Spanish business and academic circles, was increasingly replaced after 1957 by economic management being placed in the hands of technocrats charged with a relaxation of state controls and improvements in overseas trade as a means of attaining economic growth.

In 1958, the limitations of this 'new' approach had also become apparent and a full-scale deflationary and 'stabilisation' programme was launched.

> This marked the start of Spain's full acceptance into the international financial and trading communities. In return for credit restrictions, public expenditure cuts, tax increases, and an easing of restrictions on foreign investment, and some liberalisation of trade, international support grants were made available to assist in a devaluation of the peseta.[16]

It was against this background of increasing Spanish involvement in the international community that the Spanish Foreign Minister,

Señor Fernando de Castiella, paid a five-day visit to London in 1959.[17] Señor de Castiella and the Spanish Ambassador in London, the Marquess de Santa Cruz, had discussions at the Foreign Office with the British Foreign Secretary, Mr Selwyn Lloyd, and later visited Prime Minister Macmillan at Downing Street. The Foreign Office issued a statement in which the talks were described as being held 'in a cordial atmosphere' and that they had dealt with 'all matters of Anglo-Spanish relations'. No precise details about the talks were released but there was a widespread belief that they had encompassed Spain's entry into the OEEC (Organisation for European Economic Co-operation), regulations affecting the daily movement of Spaniards working in Gibraltar, and facilities for British tourists to enter Spain via Gibraltar.

This new cordiality between London and Madrid was confirmed in July 1960 when Señor Castiella paid an official visit to London at the invitation of the Government.[18] According to a joint communiqué, Señor Castiella and Mr Selwyn Lloyd 'had a cordial exchange of views and agreed on their determination to improve the relations between the two countries'. The discussions were said to have covered the world situation, including items of specific Anglo-Spanish interest, and trade relations, but had not been concerned with formal negotiations on any specific topic. A cultural convention was signed during Señor Castiella's visit which provided for an exchange of students and research workers, scientists and professional experts, and of literature. Provision was made for the establishment of a mixed commission, meeting alternately in each country, to review progress under the convention. Finally, Mr Selwyn Lloyd accepted an invitation to pay a return visit to Spain. Addressing a luncheon hosted by the Institute of Directors, Señor Castiella expressed the belief that there was 'no reason whatsoever' standing in the way of closer economic collaboration between Spain and Britain.

Priority now seemed to be given by both London and Madrid to improving Anglo-Spanish relations, and Gibraltar, a potential source of discord, had assumed a lowly position on the political agenda that was emerging from the two capitals.

It was to be Lord Home, Mr Selwyn Lloyd's successor as Foreign Secretary, who paid the three-day official return visit to Spain in 1961 after a visit of equal duration to Portugal.[19] However, despite the new atmosphere of cordiality, even during Lord Home's visit questions were raised in the House of Commons with regard to

Anglo-Spanish relations.[20] Those questions arose from remarks made by the British Home Secretary, Mr Butler, who was on holiday in Spain at the time, at a private dinner given for him by the Spanish Foreign Minister, in Madrid on 21 May.

Mr Butler was reported by the Spanish Ministry of Information as having said that it was 'shameful' that Spain had been kept out of international life for so long, especially as she represented an essential factor in opposing the communist threat to Western Europe; that he was of the opinion that Spain should be fully integrated into the Western bloc; and that both Spain and Portugal represented key elements necessary to the future greatness of Europe. Not surprisingly, Mr Butler's remarks were given considerable exposure in the Spanish media. In London, Mr Denis Healey, the Opposition spokesman on Foreign Affairs, responded to reports of Mr Butler's speech by saying that a question would be tabled in the House of Commons asking the Prime Minister to what extent Mr Butler's views represented the policy of the Government. No doubt anxious to reduce the impact of Opposition criticism of the Government of which he was a distinguished member, Mr Butler released a statement in Granada on 24 May in which he maintained that remarks attributed to him by the Spanish Ministry of Information had been 'very much exaggerated'. He observed that his remarks had been translated into 'rather more grandiose language' than he had actually used. Offering an alternative version of his comments to those reported in Spain, Mr Butler contended that he had said that it would be 'a good thing for Spain to be associated with the West' and that he had 'indicated my feelings that there should be a closer relationship between Spain and the West'; he emphasised that such sentiments were not new since they reflected the policy of the West and were important if trade, and other, relations were to flourish.

In the House of Commons on 30 May the Prime Minister answered questions put by the Opposition. Mr Macmillan conceded that the Home Secretary's remarks were 'in accord with the Government's policy of working for friendly relations with Spain and other Western countries'; that Mr Butler had made no mention of NATO nor was the question of Spanish membership of NATO mentioned by him despite reports to the contrary; that Mr Butler had not seen or agreed the text of the press release put out by the Spanish Ministry of Information; and that the full text of Mr Butler's remarks could not be published because the Home Secretary had 'lost the menu card on which he had scribbled a few notes'. The Prime Minister concluded by saying that:

> I really think that rather more has been made of this incident than it justifies. It is rather a storm in a teacup. My Right Honourable friend [Mr Butler] went to a dinner and made a few friendly observations, a summary of which was published. What the Opposition is doing is what all Oppositions do – turning this into a lurid and dramatic incident. The public is already bored with the whole thing.

Despite 'the storm in a teacup' at home, Lord Home's visit to Madrid proceeded smoothly. On his arrival the Foreign Secretary said that he had come to 'review matters of mutual interest and concern to both countries' and to return the visit paid by Señor Castiella to London in 1960. He expressed the hope that his visit would lead to 'closer co-operation and greater understanding' between Britain and Spain.

At an official banquet on 29 May Lord Home stated that:

> Anglo-Spanish relations have not been good. Nor would it be wise to ignore the fact that there are differences in emphasis in our political philosophies, which are not easily bridged. In working for improved relations we should not try to rush matters or face issues which are not ready to be settled ... If the process of building understanding is gradual, perhaps it is none the worse for that. We wish to see the friendship of our two peoples firm and lasting. If we succeed it will not only benefit our two peoples but the whole of Europe.

A communiqué released at the end of Lord Home's visit described the talks that had taken place as having been held 'in an atmosphere of sincerity and frankness' that they had covered the international situation and led to 'full agreement on both sides that unity and cohesion are essential to the strength of the Western world and its ability to overcome any crisis' and that both sides 'recognised the need for economic expansion as a means to safeguard the independence of sovereign nations'. The communiqué also stressed that Lord Home's visit had 'demonstrated the desire of the two Governments not only to improve relations between their two countries but also, through this friendship, to make a substantial contribution to the solidarity of Europe and thus to the peace of the world'.[21]

Both Governments seemed more concerned to find grounds for co-operation than to resolve those issues which might prove to contain the germs of discord; certainly Gibraltar would have fallen in the latter category. It can only be surmised that Lord Home's recommendation that, 'in working for improved relations we should not try to rush matters or force issues which are not ready to be resolved' was a clear intimation to his hosts that the present and future status of Gibraltar was a question which was not open for discussion at that time and that, were it to be raised, it carried the risk of setting back the improvement in Anglo-Spanish relations which had begun with Señor Castiella's visit to Britain in 1959.

2

THE PROBLEM GROWS

In September 1963 the question of Gibraltar's future was removed from the narrow confines of Anglo-Spanish relations and placed in the broad area of international politics when, at the request of the Spanish Government, the Colony was discussed for the first time at the United Nations by the Special Committee of 24 on the ending of colonialism. The Special Committee reached no conclusion at that time and adjourned further consideration of Gibraltar until its next session.

In the interim, two events which were to have far-reaching impact upon both Gibraltar's future and the conduct of Anglo-Spanish relations occurred. Firstly, after discussions in London between Lord Lansdowne, Minister of State for Commonwealth Relations and the Colonies, and unofficial members of the Gibraltar Legislative Council, agreement was reached on 10 April 1964 on constitutional changes for Gibraltar which entailed an increased measure of self-government for the Colony.[1] In a communiqué issued in Gibraltar on the day that the new constitutional arrangements were announced, the unofficial members of the Legislative Council made it abundantly clear that they were not seeking independence from Britain nor were they in pursuit of control over foreign and defence policy. They were, however, adamant that it was the desire of all of the people of Gibraltar that, 'Gibraltar should remain in close association with Britain'. For its part, the British Government emphasised that the new constitutional arrangements for Gibraltar corresponded to the wishes of the Gibraltarians and were in accordance with modern democratic ideas and the principles of the United Nations.

Secondly, in early June 1964 details began to emerge in the British press that agreement had been reached between Britain and Spain for a naval construction programme.[2] In a debate in the House of Commons on 16 June the Leader of the Opposition, Mr Harold Wilson, asked whether Britain had to 'sell drawings and details of frigates to a fascist country for a few million pounds' and claimed that the naval agreement would not generate much employment in British shipyards. He also asked the Foreign Secretary, Mr Butler, if he had 'received from the Franco Government a withdrawal of their claim to Gibraltar' and whether this had been made 'a condition of this arms deal'. In reply, Mr Butler emphasised that the question of Gibraltar had not been raised during the negotiations and he went on to say that, 'I only hope that this exchange in the House will not make the deal impossible'.

Mr Butler's hopes were, however, to be confounded when, on 30 June, a Spanish Government spokesman announced that Spain had broken off negotiations with Britain for the naval construction programme 'in view of attacks made on Spain by the Leader of the British Labour Party, Harold Wilson'. Such a decision could only have been taken at Cabinet level in Madrid.

The Anglo-Spanish naval construction programme was again the subject of heated debate in the House of Commons on 14 July when, under further Opposition questioning, the Prime Minister, Sir Alec Douglas-Home, stated:

> We have no reason to suppose that there would not be this order. A great many other orders would probably have followed, because the Spanish people desire to buy British . . . I have been profoundly disturbed by this development . . . These were commercial deals . . . They were deals between our shipbuilding industry, our engineering industries, the Spanish Navy and us.[3]

The Prime Minister was not slow to lay the blame for the breakdown of negotiations for the naval construction programme at the feet of the Opposition:

> A great weight of responsibility rests on the shoulders of the Leader of the Opposition and other members opposite, in that when everyone else is striving to increase exports he is striving to discourage them . . . here was a new market into which we could make inroads and the Right Honourable gentleman has effectively seen that it is closed to us . . . Opposition members must be feeling extraordinarily guilty on this matter.

Without question these two events served to sour relations between Britain and Spain after a period in which both Governments seemed to have striven to re-establish their relationship on a more cordial basis if only by an unstated agreement to avoid discussion of issues which might, if raised, have emphasised matters of divisive potential. There can, however, be no doubt that the increasing self-government afforded to Gibraltar through the Colony's new constitution and the irritation caused in Spain by criticism in Britain of the joint naval construction programme heralded a downturn in Anglo-Spanish relations; the magnitude of which can be gauged by the deliberations of the UN Special Committee of 24 when it resumed its consideration of Gibraltar in September 1964.

When the Special Committee reconvened the Spanish delegate, Señor Jaime de Pinies, contended that the policy which Britain had been pursuing with regard to Gibraltar since 1950 had been an attempt to replace the Treaty of Utrecht by moving down the road leading to self-government for Gibraltar. If the end of that road was finally reached then, Señor Pinies contended, Spain would no longer consider itself bound by the Treaty of Utrecht; which provided for Spain to be given preference to Gibraltar if Britain was to grant, sell or by any other means alienate ownership of the territory.[4] Señor Pinies insisted that Gibraltar was part of Spanish territory and that its present inhabitants had no say in its future. He pointed out that, should the Committee of 24 or the General Assembly of the United Nations decide that the most fitting way to decolonise Gibraltar was by the application of the principle of self-government to its current inhabitants, then Spain would find it impossible to maintain normal relations with the new political entity that would come into being and that, likewise, Spain could have no further contact with Gibraltar unless Britain terminated its presence there, since Spain took the view that the British grant of self-government to Gibraltar would relieve the Colony of its obligations to Britain. Señor Pinies also issued the first concise and clear warning as to future Spanish action over Gibraltar when he stated that if the British did not withdraw then Spain would consider the creation of a new political entity as simply a scheme designed to maintain colonialism and the inhabitants of Gibraltar would be regarded as personae non gratae in Spain and communications between Spain and Gibraltar would be cut. Señor Pinies suggested that Spain and Britain negotiate on the question of Gibraltar in keeping with

Paragraph 6 of the United Nations General Assembly Resolution 1514.[5]

The British delegate to the United Nations, Mr Cecil King, stated that the British Government did not accept the Spanish contention 'that Spain had any right to be consulted on changes in the constitutional status of Gibraltar', nor over the Colony's relationship with Britain, and that it was:

> satisfied that the grant of Gibraltar to Britain under the Treaty (of Utrecht) was absolute and without bar to constitutional changes in Gibraltar or to the acquisition by its inhabitants of a full measure of self-government as the Charter (of the UN) required.

Mr King voiced 'surprise and regret' at Señor Pinies' 'contemptuous and menacing references to the people of Gibraltar and the steps that Spain threatened to take against them', whilst emphasising that the British Government was 'fully conscious of their obligations to protect the welfare and defend the legitimate interests of the people of Gibraltar and would not hesitate to fulfil these obligations in whatever manner that might be necessary'. Mr King did not preclude the prospect of future change in connection with Gibraltar when he said that the British Government:

> respected the wishes and aspirations of the people of Gibraltar (and) if their elected representatives wished to advance proposals for the form of their association with Britain, [the British Government] would be ready to consider them and work out with the Gibraltarian representatives arrangements for a continuing association acceptable to both parties.

He also sought to placate the Spanish demands by offering 'an unqualified assurance that the constitutional changes recently introduced in Gibraltar [August 1965] would in no way damage the interests of Spain'.

Two petitioners from Gibraltar, Sir Joshua Hassan, the Chief Minister, and Mr Peter Isola, the Leader of the Opposition in the Legislative Council, told the Committee 'of the results of the discussions which had led to the introduction of the new . . . [1965] Constitution' and 'declared that the people of Gibraltar wished to remain closely associated with Britain, did not wish to become independent, and were opposed to being handed over to Spain against their wishes'.

The Committee's considerations ended on 16 October with the adoption of a consensus which observed that there was 'a disagreement, even a dispute, between the UK and Spain over the status and the situation of the Territory of Gibraltar'. The Committee further invited the British and Spanish Governments to engage in 'conversations in order to find . . . a negotiated settlement'. In making this recommendation the Committee was at pains to stress that 'the provision of the (UN) declaration on the granting of independence to colonial countries and peoples are fully applicable . . . to Gibraltar'.[6]

Mr King rejected the Committee's assertion 'that there was a dispute over the status of Gibraltar' and indicated that the British Government would not feel itself to be:

> bound by the terms of any recommendation of the Committee touching on questions of sovereignty. HMG did not accept that there was a conflict between the provisions of the Treaty of Utrecht and the application of the principle of self-determination to the people of Gibraltar. On the question of the future of Gibraltar they would be guided, as the UN Charter required, by what HMG regarded as the paramount interests of the people of Gibraltar.

Finally, Mr King repeated that Britain was 'not prepared to discuss with the Spanish Government sovereignty over Gibraltar' but was 'willing to discuss . . . the maintenance of good relations and the elimination of new causes of friction'.

Gibraltar thus moved from being an issue that both the British and Spanish Governments chose to leave off the political agenda to one which neither party could any longer ignore irrespective of the impact that it might have upon their relationship. As though to emphasise this point Gibraltar became a focus for action rather than for words within a day of the UN Committee of 24 consensus when, on 17 October 1964, the Spanish authorities at the frontier at La Linea began a more rigid enforcement of customs procedures and regulations against vehicles not registered in Spain than had hitherto been the case. Inevitably delays occurred. A temporary abatement came on 24 October, but frontier procedures were again more stringently enforced a week later occasioning delays of between one and ten hours to motor vehicles crossing from Spain to Gibraltar. At the same time, the export of all goods, except fresh foodstuffs, from Spain to Gibraltar was halted and visitors leaving Spain for Gibraltar were warned of impending delays. To add force to these actions the Spanish authorities ceased to renew licences for goods and passenger vehicles crossing the frontier.

Further action was taken on 21 November when the Spanish authorities began to close the frontier gates an hour earlier than previously. Inconvenience was caused to both Spanish, Gibraltarian and other citizens by this growing range of restrictions. Spanish citizens who had spent all, or the greater part, of their working lives in Gibraltar were now no longer allowed to use the passports and 'workers' passes' necessary for their daily crossing of the frontier into Gibraltar. Gibraltarians and other British citizens discovered that without warning the Spanish police and customs authorities refused to accept as valid British passports endorsed by British Consulates in Spain which stated that they had been issued or renewed 'on behalf of the Government of Gibraltar'. British passports issued or renewed in Gibraltar were also now deemed to be invalid.[7]

Inevitably, these restrictions at the frontier led to exchanges between the British and Spanish Governments in the autumn and early part of the winter of 1964. On 6 November, the British Embassy in Madrid asked the Spanish Foreign Ministry for an explanation of the new frontier restrictions only for the British Ambassador, Sir George Labouchere, to be informed by Señor Castiella, on 9 November that the frontier delays were not the deliberate result of policy. The Ambassador made further representations on 12 and 17 November and on the former date the matter was also raised by the Foreign Office with the Spanish Ambassador in London. This initial skirmishing led, on 18 November to a letter to the British Ambassador from Señor Castiella which, after referring to the consensus adopted by the UN Committee of 24, indicated a Spanish preparedness to undertake the negotiations encouraged by the Committee of 24. However, Señor Castiella's letter concluded with a warning that should a negotiated setlement not be reached then the Spanish Government would find itself obliged to revise its policy towards Gibraltar.

The British reply to Señor Castiella's letter came on 24 November. It pointed out that, although the British Government was prepared to consider, without commitment, any proposals that the Spanish Government might feel it fitting to make about discussions in keeping with the consensus of the Committee of 24, HMG was not prepared to negotiate under duress and that, therefore, it did not propose to reply to Señor Castiella's letter. The British reply also clearly indicated the inability felt by the British Government to discuss with its Spanish counterpart the question of Gibraltar whilst the frontier restrictions remained in force. It also held out the possibility that once those restrictions had been removed, and a suitable interval had elapsed, then the British Government could see its way clear to considering conversations of the kind indicated by Señor Castiella were these to be again proposed by him.

Sir Joshua Hassan and Mr Peter Isola visited London in November for talks with the Colonial Secretary, Mr Greenwood, on 24 November. Two days later Mr Greenwood told the House of Commons that he had reassured his visitors that Britain 'was fully conscious of our obligation to protect the welfare of Gibraltar and the legitimate interests of its people'.

On 10 December the British Ambassador communicated to Señor Castiella an oral message from the British Foreign Secretary, Mr Patrick Gordon Walker, who recalled that one of his first acts on taking office in the previous October had been to inform Señor Castiella of the desire of the new, Labour, British Government to maintain good relations with Spain and that such a desire still remained. In reply Señor Castiella assured his opposite number that Spain was not considering any other measures on its frontier with Gibraltar and he pointed out that the measures begun in October were those which any country was free to adopt to prevent smuggling.

By the beginning of 1965 the effect of the Spanish frontier restrictions upon the Colony had become manifest. In January 1965 only 873 vehicles crossed the frontier from Spain into Gibraltar compared to 8,691 in January 1964. The tourist industry, and businesses dependent upon it, had been severely affected and Gibraltar was having to seek alternative suppliers of goods formerly imported from Spain. In short, the exchange of Notes between the British and Spanish Governments in the latter part of 1964 had failed to produce any alleviation of the problems afflicting Gibraltar as a direct result of the introduction of frontier restrictions by Spain. In consequence, a new round of representations began on 11 January 1965 when the British Ambassador delivered a formal Note to Señor Castiella protesting against the frontier restrictions and asking for them to be immediately lifted. The Spanish Government replied on 16 January, repeating its request for direct negotiations and referring also to the problems presented to Spain by the presence of a British military base on Spanish territory. On 22 January, the British Government replied re-stating its position as outlined in the Notes of the previous months and totally rejecting the Spanish contention that Gibraltar was a British military base on Spanish soil. On the same day, in New York, the British permanent representative at the UN requested the Secretary general to circulate to all UN members the text of the two British Notes of 11 and 22 January; a tactic clearly intended to elicit support for the British position and to cast doubt upon the integrity of the Spanish Government.

British recourse to the UN initially appeared to have paid dividends and a brief softening of the Spanish attitude could be detected in a Note from the Spanish Foreign Minister of 29 January which asked if the British Government would be willing to discuss Gibraltar on the basis of the recommendations of the Committee of 24 were the Spanish Government to stop its 'defensive measures' at the frontier. The British Government reply of 8 February amounted to a reiteration of the fact that there could be no conversations between the two Governments as long as restrictions remained in force but that talks aimed at a settlement could be considered after restrictions had been lifted. However, discussions on the question of the sovereignty of Gibraltar were ruled out although recognition was afforded to the fact that constitutional changes affecting Gibraltar were a matter of interest to Spain.

Perhaps discouraged by the British Government's refusal to consider any suggestion relating to the sovereignty of Gibraltar, the Spanish Government, in a Note of 10 February, indicated a preparedness to intensify its 'defensive counter measures' unless the British Government changed the orientation of its policy and negotiated with Spain. The Note also proposed that, prior to any negotiations, Britain should restore the internal constitutional position of Gibraltar to the state in which it was before the establishment of a Legislative and Executive Council;[8] contended that the interests of Gibraltarians could be expressed through the municipality of Gibraltar; and emphasised that should the British Government accept this approach to the resolution of the problem presented by Gibraltar, then Spain would, for her part, 'ensure that no serious alterations occur in the civic life and economy of Gibraltar until the beginning of the negotiations and during the course of the same'.

This latest Spanish Note led to a statement being made in the House of Commons on 11 February by Mrs Eirene White, Colonial Under-Secretary, in which she pointed out that while it would be 'a mistake to exaggerate' the effect of the frontier restrictions, some sectors of the Gibraltarian economy, most noticeably those dependent on tourism, were being affected and that 'continuance of the restrictions would involve problems of economic adjustment'. Mrs White assured the House that as soon as Gibraltar's needs were established the Colony 'can rely on us to consider urgently and with the utmost sympathy what help we can give'.[9]

The day following her statement to the House, Mrs White, departed for a three-day visit to Gibraltar where she had talks with the Governor, Ministers and Opposition members, and senior officials and members of the Chamber of Commerce as well as meeting those who had had to leave their homes in Spain and move into accommodation provided by the authorities in Gibraltar.[10] It was only after Mrs White's statement in the House of Commons and visit to Gibraltar that the British Government replied to the Spanish Note of 10 February. In a Note of 22 February, the British Government again reiterated its position it regarded the frontier restrictions as a deliberate attempt to influence the situation appertaining to Gibraltar; an attempt which would prevent the holding of conversations anticipated by the Committee of 24; and stressed that the conditions stipulated in the Spanish Note of 10 February precluded any conversations. In two further Notes of 1 and 30 March, the British Government protested to Spain over the withdrawal of 'workers' passes' from British subjects living in the Campo area[11] and objecting to the Spanish refusal to accept the validity of certain British passports.[12]

Gibraltar featured twice more in House of Commons debates in the spring and summer of 1965. Firstly, on 1 March, Mr Michael Stewart, Foreign Secretary, answered questions in the House during the course of which he emphasised that, 'we will take every step necessary to look after the Gibraltarians' and that 'if the Spanish Government feel that they have any genuine grievances to discuss, I have made it clear that we are perfectly prepared to talk to them'. The Gibraltar question had now assumed such proportions that on 5 April the British Government published a White Paper entitled 'Gibraltar – Recent Differences with Spain' which recorded the Government's view of the dispute.[13] Secondly, the Prime Minister, Mr Harold Wilson, answering questions in the House on 29 June, stated that there had been 'no change in the Spanish position' and added that, 'the people of Gibraltar are being defended and sustained'.

The defence and sustenance of Gibraltar was already underway and indications of this had been provided earlier, on 3 May, when Sir Joshua Hassan, had announced that the Gibraltarian Government had decided to prepare a comprehensive economic survey with a view to producing a 'master plan' for the Colony's re-development; a course of action viewed as being necessary, even if the attitude of the Spanish Government were to change, because Sir Joshua believed that the ultimate solution for Gibraltar lay in the full and effective use of the Colony's own resources.

Sir Joshua revealed that a study group, involving consultants from Britain, had already been constituted and was at work. The study group's terms of reference were to produce a report which would serve as a basis for the 'master plan' by the early autumn. The study group had been requested to give priority to the following objectives: (1) a substantial increase in the provision of attractive holiday accommodation; (2) making Gibraltar a holiday resort in its own right; (3) establishing further light industries; (4) and providing additional recreational and commercial facilities required to match any increase in the number of residents and visitors.[14]

The British Prime Minister's claim that 'the people of Gibraltar are being defended and sustained' took a more tangible form during talks in London between Sir Joshua and Mrs Barbara Castle, Minister for Overseas Development, on 22 July, when it was announced that the British Government was making £1,000,000 available to Gibraltar in colonial and development grants over the following three years, together with a further £200,000 in Exchange Loans should the Government of Gibraltar have recourse to them. Mrs Castle stated in a written parliamentary reply that the one million pound grant was being made to Gibraltar in view of 'the urgent need for the Colony to reorganise its economy and to provide more housing to meet the needs of British subjects who have been obliged to leave Spain'.[15]

With no solution of the Gibraltar question in sight the Spanish Government published its version of the dispute on 4 December in a 550-page 'Red Book'; a response to the British White Paper of April. The Red Book stated that, where Spanish diplomatic efforts to come to an understanding with Britain were concerned, it had been thought in Madrid that the visit paid to London by Señor Castiella in 1960 and the return visit paid to Spain by Lord Home in 1961 had opened a dialogue between the two countries. The Red Book claimed, however, that 'the British refusal to eliminate the real causes of one or two problems ruled out the possibility of that dialogue and left it frustrated'. An exclusive Anglo-Spanish settlement seemed as far away as ever.

3

A DEEPENING PROBLEM

1965 ended with Gibraltar again being the subject of discussion at the United Nations. On 16 December a number of resolutions on the ending of colonialism was approved by the Twentieth General Assembly arising from reports of the Special Committee on the Granting of Independence to Colonial Countries and Peoples (Committee of 24) and which had been approved previously in the Assembly's Fourth Committee (Trusteeship and Non Self-Governing Territories). Where Gibraltar was concerned the British and Spanish Governments were invited to begin negotiations on the Colony without delay and to keep the UN informed of the course and outcome of their discussions.[1]

Following the UN General Assembly resolution the British and Spanish Governments agreed to enter into discussions on Gibraltar. The first, of what was to be four rounds of talks, began at the Foreign Office in London on 18–20 May 1966. Before the first round of talks commenced, Sir Joshua Hassan visited London for discussions with British Ministers and prior to returning to Gibraltar he stressed that the Gibraltarians desired to remain British and emphasised that 'we would never accept that the British Government has the right to change our status against Gibraltar's wishes'.

A strong team of Ministers was fielded by both Governments at the first round of talks. The Foreign Secretary, Mr Michael Stewart, headed the British delegation, whilst his opposite number in the Spanish Government, Dr Fernando Castiella, led the visiting party. The Spanish delegation made a formal claim to sovereignty over Gibraltar in a four-point plan:

1 The signing of an Anglo-Spanish convention, the first Article of which should provide for the cancellation of Article 10 of the

23

Treaty of Utrecht of 1713 and 'the restoration of the national unity and territorial integrity of Spain through the reversion of Gibraltar'.

2 Spain would accept the presence at Gibraltar of a British military base 'whose structure, legal situation, and co-ordination with the defence organisation of Spain or the free world would be the subject of a special agreement attached to the convention'.

3 A legal regime protecting the interests of the present citizens of Gibraltar should be the subject of an additional Anglo-Spanish agreement, which would be registered with the UN. In that agreement, in addition to the appropriate economic and administrative formulae, a personal statute would be established by which, among other fundamental rights – such as freedom of religion – the British nationality of the present inhabitants of Gibraltar would be respected and their right of residence guaranteed, as would the free exercise of their lawful activities and a guarantee of permanence in their place of work.

4 The convention should take effect after the two additional agreements referred to in (2) and (3) above had been registered with the UN.

Mr Stewart replied to these proposals on behalf of the British Government. He stressed that Britain had no doubts as to her legal rights in Gibraltar under the Treaty of Utrecht and again emphasised that the sovereignty of Gibraltar was not a matter for negotiation. In Gibraltar, Mr Stewart's remarks were welcomed by Sir Joshua Hassan who stated that all Gibraltarians would be 'intensely thankful for Mr Stewart's declaration' since it unequivocally specified that there was 'no question of Britain handing over the Rock'.

The second round of talks, held in London on 12–13 July 1966, saw less prestigious delegations assuming the task of negotiating the future status of Gibraltar. The Spanish delegation was led by the Marquis de Santa Cruz, Ambassador to Britain, and the British delegation by Mr H.A.F. Hohler, a senior official at the Foreign Office. On this occasion it was the turn of the British delegation to advance four proposals:

1 The internal self-government granted to Gibraltar should be exercised by municipal authorities instead of by an Executive Council and a Legislative Council. Accordingly, the Gibraltar Council of Ministers would become a Municipal Council, and the Chief Minister and the other ministers would become Mayor

and Aldermen respectively.[2]

2 Spain would be offered facilities at the airport, both for civil and military planes, and in the dockyard.

3 The Spanish Government would appoint a representative in Gibraltar to oversee Spanish interests there and to fulfil consular functions.[3]

4 Joint arrangements should be made to combat smuggling.[4]

If nothing else, the two rounds of direct talks between the British and Spanish Governments over the future of Gibraltar had ended the exchange of Notes which had been so prominent a feature of the previous year and facilitated a clear statement of the respective positions that each Government had adopted over the Colony's future. However, no further signs of progress, nor any steps towards the resolution of the problem posed by Gibraltar to Anglo-Spanish relations could be discerned and it was, therefore, not surprising that the third round of talks made no further headway.[5]

The climate for the third round of talks had been dampened prior to their commencement when Spain had forbidden British military aircraft en route to Gibraltar and other destinations to fly over Spanish territory.[6] A similar discordant note was to precede the fourth round of talks when, on 5 October, the Spanish Government issued a decree providing for the closure of the customs post at La Linea, the entrance to Gibraltar from mainland Spain, on the grounds that the traffic in merchandise was 'virtually non-existent'.

On the day following this latest Spanish decree it was announced in London that the British delegation at the fourth round of talks would propose that legal issues in dispute between Britain and Spain should be referred to the International Court of Justice. Unquestionably, the British Government had proposed this course of action as a direct response to the Spanish decision to close the customs office at La Linea; a motive made clear in a statement issued by Mr George Brown, Mr Michael Stewart's successor as Foreign Secretary, in which he expressed disappointment that the Spanish Government had 'failed to respond to our representations on the new restrictions . . . on movement between Spain and Gibraltar' and regretted that the Spanish Government 'should have sought to apply pressure in this way on the eve of the resumption of talks . . . [which] should be given every chance of success'. Mr Brown left his audience in no doubt about the direct link between events at La Linea and the British proposal:

> when the talks are resumed the British delegates will propose that the legal issues in dispute . . . should be referred to the International Court of Justice. HM Government considers that this proposal is a more appropriate method of conducting this dispute than imposing restrictions on movement across the frontier at La Linea.

Accordingly, when the fourth round of the Anglo-Spanish talks of 1966 began in London on 10–11 October the British delegation puts its proposal to its Spanish counterpart. There was no immediate Spanish reply but one was later forthcoming in a Note presented in London on 14 December in which the Spanish Government rejected recourse to the International Court of Justice because it would contradict 'the resolution of the United Nations recommending that the colonial situation in Gibraltar be brought to an end and regretting any delay which might arise in the process of decolonisation'.[7] In the interim between the ending of the fourth round of talks and the Spanish Note of 14 December the Spanish Government had alleged repeated violations of Spanish air space by RAF planes in the Gibraltar area.[8]

The Spanish Government's decree of 5 October came into effect on 24 October when the frontier at La Linea was closed to all but pedestrians. This ban on all vehicular traffic to and from Gibraltar meant that an estimated 7,000 Spaniards who daily entered the Colony from Spain had to walk across the frontier at La Linea. The absolute ban on motor traffic occasioned the Governor of Gibraltar, Sir Gerald Lathbury, to warn that the Rock was likely to face 'an indefinite period of savage blockade'. The Spanish tightened the net they had cast around Gibraltar on 12 November when the Spanish frontier authorities at La Linea announced that they would only accept UK passports issued by the Foreign Office in London and that passports issued by the Governor of Gibraltar were no longer valid. This meant that Gibraltarians were effectively barred from entering Spain.

The Gibraltar situation had now developed to such a degree of severity that it warranted further debate in the House of Commons especially as, on 30 October, windows in the British Consulate in Barcelona were smashed during a demonstration staged by some 200 young Falangists demanding 'the restoration of Gibraltar to Spain'.[9]

Addressing the House, on 31 October Mr Brown reaffirmed that 'Gibraltar is of course British by right'. He went on to tell the House that:

> the recent Spanish actions, including the Spanish Government's activity in imposing new restrictions at the frontier and mob attacks on our Consulates ... cannot affect this fact ... If the Spanish Government accept the proposal (to refer Gibraltar to the International Court of Justice) it will be for the Court to decide on all questions referred to it, including the question of sovereignty ... it remains the firm intention of the Government to sustain Gibraltar in her present difficulties and if we think that further financial aid is needed for that purpose we will provide it.

Gibraltar certainly needed both sustenance and 'further financial aid' to help stabilise an economy severely mauled by the Spanish imposition of frontier restrictions. To assess Gibraltar's needs, Mr Frederick Lee, Secretary of State for the Colonies, visited the Rock in early November and whilst there announced that the British Government would make an immediate grant of £600,000 to Gibraltar to enable the Government of Gibraltar to begin a development programme covering the period up to 1970 and designed to make the Gibraltarian economy self-sufficient and totally independent of that of Spain.

With an economy lacking the infra-structure necessary for self-sufficiency and subject to the damaging vagaries of Spanish action many in Gibraltar saw their future as dependent upon a decisive move by the British Government which would inspire confidence and dispel uncertainty. Such a belief was contained in a letter to the British Prime Minister, Mr Harold Wilson, from Major Robert Peliza, Chairman of the Integration with Britain Movement in Gibraltar, which emphasised that 'the final settlement of the constitutional relationship with Britain will dispel the Gibraltarians' fear of losing their right to remain British in British Gibraltar'. Major Peliza drew the Prime Minister's attention to the Act of Union offered by Britain to both Malta and Rhodesia and added that 'the Movement feels sure that should the Gibraltarians, on de-colonisation, choose by referendum to unite indissolubly with GB, HM Government will show the same generosity to Gibraltar as they have to Rhodesia'.

As 1966 drew to a close, Gibraltar again assumed significance at the United Nations. At its 21st session in New York, the General Assembly, on 20 December, passed a resolution on Gibraltar, by 101 votes to 0 with 14 abstentions and 7 countries absent, which, after observing the 'willingness of the Administering Power and the Government of Spain to continue its present negotiations' and regretting 'the occurrence of certain acts which had prejudiced the smooth progress of the negotiations':

1 regretted 'the delay in the process of decolonisation';
2 called upon Britain and Spain 'to continue their negotiations taking into account the interests of the people of the territory';
3 requested Britain 'to expedite, without any hindrance and in consultation with the Government of Spain, the decolonisation of Gibraltar';
4 asked that a report should be made to the UN Special Committee on colonialism (the Committee of 24) 'as soon as possible, and in any case before the 22nd session of the General Assembly'.[10]

The General Assembly resolution on Gibraltar had already been passed by its Trusteeship Committee on 17 December by 78 votes to 0 with 12 abstentions. At his own request, Sir Joshua Hassan had addressed the Trusteeship Committee during its debate on Gibraltar and had told it that the people of Gibraltar enjoyed a full measure of self-government; that political parties of all shades of political opinion operated freely; that the process of decolonisation had started in 1950 with the establishment of internal self-government through the creation of Executive and Legislative Councils; and that the Gibraltarians wished to remain associated with Britain.[11]

Both Spain and Britain felt able to vote in favour of the General Assembly resolution on Gibraltar. The Spanish delegate, Señor Don Manuel Aznar, felt that the resolution contained 'constructive elements' whilst the British delegate, Lord Caradon, held that it did not prejudice 'the question of the type of decolonisation which would best fit the circumstances of Gibraltar'. Lord Caradon drew specific attention to the phrase 'taking into account the interests of the people of the territory' and emphasised that decolonisation could 'never mean the incorporation of Gibraltar into Spain against the wishes of the people'.

In much the same way that 1965 had been characterised by a frequent exchange of Notes between the British and Spanish Governments over the question of Gibraltar, 1966 was characterised by rounds of talks between them. Yet neither approach had led to a settlement of the Gibraltar question despite resolutions passed at the United Nations that an end to the dispute be reached by negotiation.

However, the United Nations resolution of December 1966 could not be ignored for it had clearly placed the onus on the British Government, as the colonial power, to initiate further discussions with the Spanish Government which would lead to the eventual decolonisation of Gibraltar. Accordingly, on 29 March 1967, a British Note was handed to the Spanish Ambassador in London which proposed the resumption of talks between the two Governments on the Gibraltar issue. Subsequently, the Foreign Office announced that talks would begin in London on 18 April. This British initiative clearly complied with the General Assembly resolution of the previous December. In the interim, however, London had been the venue for talks between Mrs Judith Hart, Minister of State at the Commonwealth Office, Sir Joshua Hassan and Mr Peter Isola of Gibraltar and the Colony's Governor, Sir Gerald Lathbury. Although specific details of what passed between Mrs Hart and the Gibraltarian representatives was not revealed, the Commonwealth Office stated that 'complete agreement' had been reached.

If such a phrase truly encapsuled the outcome of those talks then only a limited number of possibilities as to the content which provided the basis for such a consensus are feasible. Presumably, the British Government reassured the Gibraltarian representatives that in its opinion Gibraltar was British by right, that Britain would sustain the Colony during its present difficulties and would offer such financial aid as was deemed necessary to ensure the Rock's continued survival and economic well-being. No doubt mention was made also of the British Government's resolve to remain true to that part of the General Assembly resolution which stressed that 'negotiations [with Spain must take] into account the interests of the people of the territory'. In short, the Gibraltarians must have been assured that Britain would not enter into any agreements with Spain, as a result of any future negotiations between London and Madrid, unless the content of those agreements met with the full, unqualified and freely expressed approval of the Gibraltarians themselves.

Again, however, the course of Anglo-Spanish negotiations was not to begin in a climate supportive of success. In very much the same way that it had imposed restrictions on the use of the land frontier at La Linea less than a week before the round of Anglo-Spanish talks on

Gibraltar held in London in October 1966, the Spanish Government, on 12 April, and again less than a week before the start of the new talks, issued a decree prohibiting all flights by foreign aircraft, both civilian and military, over an area around Algeciras immediately contiguous to Gibraltar, extending over an 80-mile stretch of coast between Tarifa (south west of Gibraltar) and Estepona (north east of Gibraltar) and for some distance inland. Imposed for what the Spanish Government described as 'national security reasons' the new ban on flights extended also to adjacent Spanish territorial waters. A further justification offered by Madrid for the ban was the allegation that 'repeated violations' had been made of Spanish airspace by British military aircraft using Gibraltar and that, in consequence, the Spanish Government had been obliged to introduce the restrictions under Article 9 of the Chicago Convention of 1944; that notification had been given to the International Civil Aviation Organisation (ICAO); and that the ban would come into effect at the expiration of one month's notification to the ICAO as required by the organisation's rules.[12]

Answering questions in the House of Commons on 13 April, Mr Bowden, Commonwealth Secretary, said that 'we of course uphold to the full our right to use the airfield at Gibraltar'. Mr Bowden 'deplored' the new restrictions intended by Spain and announced that in the light of them the Anglo-Spanish talks scheduled for 18 April had been postponed. He also announced that both civilian and military flights to Gibraltar would continue.[13]

The question posed over the continued utilisation of Gibraltar's airfield by British planes as a result of the latest Spanish restrictions was a far more tangible one than that posed over the sovereignty of the Colony and its decolonisation; but one which, nevertheless, had effectively risen to prevent discussion of those more substantial, if more ephemeral, issues. Being more tangible it inevitably became the focus of attention in the following weeks. In a Note presented in Madrid on 24 April the British Government refuted Spain's persistent allegations of violation of Spanish airspace by British military planes at Gibraltar.[14] On the same day in London the Spanish Ambassador was informed that the British Government had had recourse to the Council of the ICAO and had requested the Council to place the imposition of air restrictions by Spain at Gibraltar on the agenda for its next meeting.[15] Much to British chagrin the ICAO Council declined to offer a ruling on the dispute and, accordingly, the Spanish ban on flights across its territory to and from Gibraltar became effective at midnight on 14–15 May.[16] On 17 May, the British Ambassador to Spain, Sir Alan Williams, called at the Spanish Foreign Ministry with a Note proposing direct Anglo-Spanish negotiations on the Spanish ban on flights near Gibraltar. The Note suggested that talks should begin, either in London or Madrid, on or soon after 25 May and focus on 'the effects of the prohibited area on the use of Gibraltar by civil aircraft'. Sir Alan received an immediate response, described in Madrid as a Note *verbale*, in which the Spanish Government called for the continuation of negotiations for the decolonisation of Gibraltar 'in accordance with the UN General Assembly resolution of December 1966'. The Note *verbale* contended that, in the view of the Spanish Government, the problem posed by Gibraltar could only be resolved 'within the framework of negotiations for the decolonisation of Gibraltar'; in consequence, Spain asked that such negotiations should be resumed 'urgently and without delay'. Further Notes were exchanged between the two Governments on 18, 22 and 26 May and Anglo-Spanish talks began in Madrid on 5 June. The talks, however, made no progress and eventually broke down on 8 June. The official reason given for the impasse was that Spanish insistence that purely technical talks between the two countries over access to Gibraltar airfield, as desired by Britain, could only take place subject to Britain ceasing to contest Spanish sovereignty in the area of the airfield and ending military flights both to and from it.

The Spanish air restrictions, following those imposed by the Spanish authorities on the ground at La Linea, were of grave concern to the Government of Gibraltar and, on 20 May, Sir Joshua Hassan and members of the Legislative Council appealed to the UN Secretary-General, U Thant, to ask him to request the Spanish Government to end its aerial restrictions around Gibraltar. The appeal noted the UN General Assembly resolution on Gibraltar of December 1966, and especially that part of it which said that the interests of the people of Gibraltar must be taken into account in determining the Colony's future, before claiming that the Spanish restrictions on air traffic were intended to bring about the economic bankruptcy of the Colony and were thus occasioning great hardship to its population.

The increasing severity of Spanish restrictions on Gibraltar, both by land and by air; the UN resolution placing responsibility on Great Britain, as the Colonial power, to take the initiative in decolonising Gibraltar; and the failure of Anglo-Spanish talks to resolve the question of the Rock's future status, led the British Government to embark upon what was seen by many as a radical step. On 14 June, Mrs Judith Hart told the House of Commons that a referendum would be held in Gibraltar, probably early in September, in which the Gibraltarians would be given the choice of continuing their association with Britain or of opting to pass under Spanish sovereignty:

> the people of Gibraltar should be invited to say which of the two following alternative courses would best serve their interests:
> a) to pass under Spanish sovereignty in accordance with the terms proposed by the Spanish Government to HM Government on 18 May 1966
> b) voluntarily to retain their link with Britain, with democratic local institutions and with Britain retaining its present responsibilities.

The announcement of the referendum produced a predictable and strongly critical Spanish reaction which was contained in a Note presented, by the Spanish Ambassador in London, to the Foreign Office on 3 July. The Spanish Note emphasised that previous UN resolutions had stipulated that Britain and Spain should resolve their differences over Gibraltar by direct negotiation. The Spanish Note further contended that should Britain proceed with its 'unilateral decision' to hold the referendum it would be embarking on a path 'contrary to that indicated by the UN, with all of the consequences which that implies . . . the questions that are to be put at the projected referendum contravene in their substance the UN recommendations concerning the manner in which an end must be put to the colonial situation in Gibraltar'.

On the same day that the Spanish Note of protest was delivered to the Foreign Office an Order in Council was issued in London for the holding of the Gibraltar referendum. This Order coincided with the receipt of a second Spanish Note on the referendum which reiterated the arguments of its predecessor and was also sent to the UN. This second Note protested that the referendum was illegal because of its violation of General Assembly resolutions, in that it was being held without consultation with the Spanish Government and because it contravened the Treaty of Utrecht. The British Government was not, however, to be swayed from the path upon which it had embarked and, on 18 August, Mr Arnold Smith, Commonwealth

Secretary-General, announced that a small, independent team of observers drawn from Commonwealth countries would monitor the referendum at the request of the British Secretary of State for Commonwealth Affairs and would report its findings to the British Government.[17]

The focus of attention now switched back to the United Nations in New York where, on 1 September, the General Assembly's Special Committee on Colonialism, the Committee of 24, supported a resolution opposing the holding of a referendum in Gibraltar and calling upon the British and Spanish Governments to resume direct negotiations on the future of Gibraltar.[18]

The Committee of 24's resolution was a blow to the British Government which, judged by the remarks of its representative at the UN, Lord Caradon, had not been envisaged. Lord Caradon described the Committee's resolution as a 'wholly partisan document' and went on to indicate that it stood in contravention of the resolution adopted at the last General Assembly meeting which had stipulated that the interests of the people of Gibraltar should be taken into account. He stated that the referendum would take place as planned, despite the Committee's latest resolution claiming that it contradicted previous UN resolutions on Gibraltar, before emphasising that the referendum was in accordance with the provision in the UN Charter which held that the interests of the inhabitants of a non self-governing state were paramount.

When news of the Committee's resolution reached Gibraltar it drew an immediate response from Sir Joshua Hassan who sent a message to the UN Secretary-General in which he described as 'inconceivable' the fact that a UN Committee should have passed a resolution 'which appears to disregard the interests of a colonial people and would seek to deny them the elementary human right of stating their interests through a referendum'.

Spanish and UN opposition to the holding of the referendum generated hostility in Gibraltar where the majority of citizens shared Sir Joshua Hassan's view that the failure to hold the referendum would deprive them of a say in their own future. The opposition to the holding of the referendum also served to further cement the close ties which linked the people of the Rock to Britain.

As Lord Caradon had pledged at the UN, the referendum took place in Gibraltar on 10 September and it came as no surprise when it resulted in an overwhelming pro-British vote. As Mrs Hart had previously specified, the voters of Gibraltar were given a choice

33

between passing under Spanish sovereignty 'in accordance with the terms proposed by the Spanish Government on May 18, 1966' or of 'voluntarily retaining the link with the UK, with democratic local institutions and with the UK retaining its present responsibilities'. Of the 12,762 registered voters, the right to vote being exercised only by persons normally resident in Gibraltar, 12,138 voted for the continuation of the association with Britain and only 44 voted to pass under Spanish sovereignty in a turn-out of 95.46 per cent.

In London, on 6 October, the report of the Commonwealth observer team, which had monitored the conduct of the referendum in Gibraltar, was published. The report stated that the administrative arrangements for the referendum had been executed in a 'fair and proper manner' and that 'adequate facilities for the people in Gibraltar to freely express their views' had been provided. The observer team had spent eleven days in Gibraltar and had visited all of the polling stations on the day of the referendum. The unanimous view of the team was that the conduct of the referendum had fully conformed to the 'requirements for the free expression of choice through the medium of the secret ballot'.

Even at the eleventh hour the Spanish Government had sought to render the outcome of the referendum invalid when, on 6 September, a Note had been passed to Britain requesting that Anglo-Spanish negotiations be resumed in line with the decolonisation of Gibraltar contained in the resolution of the UN Committee of 24 of 1 September.

4

NO SOLUTION IN SIGHT

The Gibraltarian electorate had delivered its verdict on the future status of the Rock in the referendum of 10 September 1967 and by so doing had produced an impasse in Anglo-Spanish relations. The Spanish Government could have recourse to the support of the General Assembly in the UN and its relevant committees in its bid to secure sovereignty over Gibraltar whilst the British Government had the buttress of the clearly expressed wishes of the overwhelming majority of the Gibraltarian electorate in its favour in seeking to maintain the Colony's close links with the UK. However, despite the weight of argument and legal and moral certitude which both governments could bring to bear on the future status of Gibraltar, it must have been apparent in both London and Madrid that the problem posed by Gibraltar was no nearer an agreed solution nor was it a problem which would simply go away. Furthermore, Gibraltar was clearly discernible as a factor with the potential to prejudice the wider context of Anglo-Spanish relations.

Perhaps in view of this seemingly irreconcilable impasse and with the outcome of the referendum in Gibraltar now known, the British Government responded positively to the Spanish Note of 6 September 1967. The British Foreign Secretary, Mr George Brown, handed a reply to the Spanish Ambassador on 20 October suggesting that Anglo-Spanish negotiations over Gibraltar be resumed in Madrid towards the end of November; at which the British Government would be represented by Mr John Beith, Assistant Under-Secretary at the Foreign Office. It was emphasised in London that the talks, if resumed, should focus upon two interdependent issues: the lifting of the Spanish land and air restrictions imposed upon Gibraltar and the improvement of relations between Spain and Gibraltar. However, the British suggestion of new talks made clear that HM Government had no intention of discussing Spain's claim to sovereignty over Gibraltar given the vote of support elicited from the Gibraltarians themselves for maintaining their links with Britain in the recent referendum. Such a proposal proved unacceptable to the Spanish Government which rejected the British offer of talks in a Note of 28 October and indicated the inability of the Spanish Government to discuss the question of Gibraltar unless the British Government invalidated the referendum result and resumed negotiations in accordance with the resolution passed by the UN Committee of 24 on 1 September.

The British were, however, resolute in their attempts to resume Anglo-Spanish talks on Gibraltar and placed a further proposal before the Spanish Government on 22 November which suggested that Mr Beith visit Madrid on 30 November for talks. This proposal was rejected by the Spanish Government on 23 November which said that it would prefer to delay debate until 10 January 1968; in other words until the UN General Assembly had again debated the question of Gibraltar at its 22nd Session. Clearly, the Spanish Government felt that the weight of international opinion was balanced in its favour but, equally clearly, it was not prepared to leave the outcome of international deliberations to chance and during the course of the General Assembly's Trusteeship Committee debates on Gibraltar, which began on 7 December, it issued a second Red Book which outlined the negotiations with Britain since May 1966. Significantly, the second Red Book described the recent referendum held in Gibraltar as 'still another side tracking manoeuvre which world opinion discounted' and declared that Spain did not accept that the future sovereignty of the Colony depended upon the wishes of its inhabitants.

When the Trusteeship Committee concluded its debate on 16 December it supported a resolution which (1) regretted the interruption of the Anglo-Spanish negotiations, (2) described the holding of the Gibraltar referendum of 10 September as a 'contravention' of the General Assembly's 1966 resolution and of the resolution passed by the Committee of 24 on 1 September (3) invited the Spanish and British Governments 'to resume without delay' negotiations 'with a view to putting an end to the colonial situation in Gibraltar and to safeguarding the interests of the population upon the termination of that situation' and (4) requested the UN Secretary-General 'to assist' the two Governments and to report to the General Assembly's 23rd Session.[1]

During the course of its deliberations the Committee listened to representations made by Sir Joshua Hassan and Mr Peter Isola, and the Mayor of San Roque as well as to those made by a Gibraltarian trade union leader and a member of the Spanish Cortes. The Trusteeship Committee's resolution was passed to the General Assembly where it was endorsed by 73 votes to 19 with 27 abstentions on 19 December.[2] The British delegate, Lord Caradon, described the resolution as being 'unworthy of the UN and a disgrace to the Committee'; a description which very much reflected the tenor of his comments on the Committee's earlier resolution of 1 September 1967.

Commenting upon the General Assembly's support for its Committee's resolution Sir Joshua Hassan was scathing in his observation that 'abuse of fact, distortion and deliberate lies have won the day'. He declared that Gibraltar was ready to resist being handed to Spain and that Gibraltarians 'are not going to be deterred by any more restrictions from the Spanish'.

With the UN again giving support to the Spanish stance on Gibraltar it was not surprising that the adoption of the latest resolution found favour in Madrid where it was described as giving the Spanish Government 'a glittering victory [in] the diplomatic battle of Gibraltar'.[3] Heartened by the weight of overt, international support the Spanish Government now took the initiative in proposing, on 12 January 1968, that early talks be held between Madrid and London on the return of Gibraltar to Spanish sovereignty. With the diplomatic initiative again wrested from it by the most recent UN resolution, the Anglo-Gibraltar camp now set about clarifying its stance. A Gibraltarian ministerial committee had already been engaged in preparing proposals relating to the Colony's future constitutional position and it was for talks with members of that committee and others that Lord Shepherd, Minister of State at the Commonwealth Office, visited Gibraltar on 4–10 February. The Spanish Government again took exception to the bilateral nature of these talks and recorded its opposition to them in a Note handed to the British Ambassador by Señor Castiella which renewed the Spanish call for Anglo-Spanish discussions on the transference of Gibraltar to Spanish sovereignty.

Although undoubtedly opposed to the direction that international opinion, as symbolised by the UN resolution, was moving on the question of Gibraltar, the British Government could not discount the weight of that opinion and, therefore, proposed to its Spanish counterpart, in a Note of 19 February, that talks be resumed on Gibraltar's future status. Significantly, the British proposed that each party to the discussions should be free to raise any subject that it wished; in short, an open agenda was being suggested. Whilst agreeing to the talks, the Spanish Government did not find such flexibility suited to its purposes. Having gained support at the UN, Madrid now sought to capitalise upon the advantage which favourable international opinion afforded and stated a preference, which the British Government declined to countenance, for focusing the proposed talks exclusively on the UN General Assembly's resolution of 19 December 1967. Britain's refusal to accept this alternative format for Anglo-Spanish negotiations on Gibraltar was stated by the Spanish Government as being 'designed to maintain the territory's colonial status' and it was pointed out that in those circumstances Britain could not reasonably expect Spain to be a party to the perpetuation of a 'colonial situation on her soil' by granting new facilities or by restoring ones not specified in the Treaty of Utrecht.

Nevertheless, a new round of Anglo-Spanish talks opened in Madrid on 18 March but broke down two days later. A Spanish communiqué, issued at the end of the talks, bluntly stated that no 'positive results' had been achieved because of the refusal of the British representative, Mr John Beith, to discuss 'compliance' with the UN resolution. The Spanish communiqué further contended that this refusal by Mr Beith had resulted in him failing 'in his attempt to obtain from Spain facilities to consolidate the British military and colonial presence on the Rock'. Presumably it was Spanish frustration at the inability of its delegation to capitalise upon favourable international opinion and to wring concessions from the British negotiators which led the Spanish Minister of Information, Dr Manuel Fraga Iribarne, to say, on 22 March, that since Britain intended to maintain 'the present colonial status' of Gibraltar on the basis of Article 10 of the Treaty of Utrecht, the Spanish Government had decided 'to apply progressively the clauses of the Treaty which guaranteed Spain's defence against any possible attempt at expansion of the Gibraltar colonial situation'. What 'expansion' Madrid envisaged was not specified.

Predictably, the British view on the failure of this latest round of Anglo-Spanish talks to make headway differed from that offered by the Spanish. The Foreign Secretary, Mr Michael Stewart, told the House of Commons on 1 April that Mr Beith had made it clear that

Britain's primary concern was with the interests of the Gibraltarians and that the breakdown of the talks had been occasioned by the Spanish delegation insisting that the discussions be exclusively based on the UN resolution.

The failure of the talks to produce any positive signs of progress gave clear testimony to a continuation of the impasse that had characterised Anglo-Spanish negotiations over Gibraltar in the previous year. However, a new factor was now introduced by a letter in the *Gibraltar Chronicle* from a group of Gibraltarian lawyers and businessmen, which had recently discussed the possibility of an agreement being reached on the Colony's future between Britain and Spain with the Spanish Foreign Minister, Señor Castiella.[4] The group signing itself the 'Doves', had suggested that a 'positive solution to the Gibraltar question' should be contained in a 'contemporary Anglo-Spanish Treaty in substitution for the outmoded Treaty of Utrecht'. The Doves made clear that they were not implying that their proposals should entail a transfer of Gibraltar's sovereignty from Britain to Spain but that they were in favour of both the Spanish and Union flags flying alongside each other in the Colony.[5] This initiative, the first visible dent in the unity which had characterised the relationship between Gibraltar and Britain, was, however, quickly revealed to be one with which many in the Colony found it impossible to sympathise. On 6 April, a deputation of Gibraltarians protested to the Colony's Governor about the letter and a crowd attacked property belonging to the Doves during the course of which a pleasure launch was burned, vehicles overturned and the windows of a jeweller's shop were broken. Despite an appeal for calm and a return to law and order, made by Sir Joshua Hassan, demonstrations continued throughout the day and troops were eventually called in by the Governor to help the local police control the crowd and prevent further destruction of property belonging to the Doves.[6]

The statement issued by Dr Manuel Fraga Iribarne, immediately after the breakdown of the Anglo-Spanish talks in March, took tangible form on 4 May, when the Spanish Government announced that from midnight on 5 May, the land frontier between Spain and Gibraltar at La Linea would be closed to all traffic, including pedestrians, except for permanent residents of Gibraltar and the five thousand Spaniards who held permits to work in the Colony. In effect this meant that everyone, except those otherwise exempted, wishing to enter Gibraltar from the Spanish mainland would have to do so by the ferry from Algeciras. Sir Joshua Hassan and Mr Peter Isola, Deputy Chief Minister, immediately departed for London for talks with Mr George Thomson, Commonwealth Secretary, which focused on the effect that the new frontier restrictions would have on the Colony's economy. Also covered were matters relating to new constitutional provisions for the Rock. The two Gibraltarian leaders remained in London for a week for further talks.

The increasing tightening of frontier restrictions by the Spanish authorities led to a wide-ranging emergency debate on Gibraltar in the House of Commons on 7 May, during which consideration was given to ways of helping the Colony's economy. A number of possible courses of action were debated covering: direct financial aid and the development of the Rock's tourist potential; retaliatory action against Spain and the recall of the British Ambassador; tighter immigration controls on Spanish workers seeking to enter Britain; and the replacement of Gibraltar's current Spanish workforce by workers brought in from other locations. Additionally, consideration was given to the measures likely to be adopted on Gibraltar by the UN later in the year and the need for coalition building among Britain's friends at the UN was recognised. Speaking for the Government, Mr Thomson said that:

> HMG will never betray the rights of the people of Gibraltar to determine where their own interests lie. I give the assurance that in no circumstances will Britain surrender sovereignty over Gibraltar against the wishes of the people of Gibraltar. We will protect and support them whatever the threats that are brought to bear on them.

The following day the Foreign Secretary, Mr Michael Stewart, informed the Spanish Ambassador in London that the British Government regarded the new frontier restrictions imposed at Gibraltar 'as arbitrary and without justification' and that the latest Spanish action had served only to make the problem posed by Gibraltar more intractable and its resolution more remote. Later that same day the Spanish Embassy stated that its Ambassador, the Marques de Santa Cruz, had delivered a Note to Mr Stewart protesting further alleged violations of Spanish airspace around Gibraltar by British military aircraft and also objecting to the 'tendentious form' in which the latest Spanish frontier restrictions had been presented to the public in both Britain and Gibraltar. The Note maintained that the frontier restrictions were not directed against the people of Gibraltar but were a result of the British Government's 'refusal to comply with the UN resolution' and its avowed intention to proceed with a constitutional conference on Gibraltar from which Spain would be excluded.

In the light of the new Spanish frontier restrictions the economic well-being of Gibraltar was a cause of growing concern both in the Colony itself and in London and it was to assess the magnitude of the damage being done to the Rock's economy, and to find ways of sustaining the Colony, that Mr Thomson visited Gibraltar on 22–4 May for further talks with the Governor, Sir Gerald Lathbury, and Sir Joshua Hassan.

On his arrival Mr Thomson was presented with a memorandum which covered the following points:

1 Gibraltar should cease to be a Colony and be accorded a new political status which would contain a reference to Britain's permanent and exclusive sovereignty.
2 Gibraltarians should be exempted from the provisions of the Commonwealth Immigration Act of 1962.
3 Gibraltar's affairs should be transferred from the Commonwealth Office to the Home Office.
4 Britain should declare an intention not to exercise her rights under the option clause of the Treaty of Utrecht without the consent of the people of Gibraltar expressed through a two-thirds majority in a referendum.[7]

Clearly, if the points made in the memorandum were to be implemented they would have had the effect of forging a still closer link between the Colony and Britain and of precluding, for the foreseeable future, any transference of sovereignty to Spain.

During his visit to the Rock, Mr Thomson announced the appointment of a mission to review Gibraltar's labour requirements with particular attention to be addressed to the position of Spanish workers daily entering the Colony. He also gave an assurance that, within the quota of vouchers available to the whole Commonwealth under the Immigration Act of 1962, 'all Gibraltarians who want to come to Britain will be able to do so'. At a later press conference, Mr Thomson said that agreement had been reached with representatives of the Gibraltar Government and that Lord Shepherd would visit Gibraltar towards the middle of July for further discussion on the Colony's constitutional status. Both of these pledges were effected in July. On 11 July, Mr Thomson told the House of Commons that Lord Beeching would chair a mission to 'review Gibraltar's manpower requirements' and that he would be assisted by Lord Delacourt-Smith, General Secretary of the Post Office Engineering Union. In Gibraltar, on 16–24 July, Lord Shepherd led a British delegation for constitutional talks with a Gibraltarian delegation which included Sir Joshua Hassan, Mr Peter Isola and elected members of the Legislative and City Councils and representatives of the Integration With Britain Party led by Major Robert Peliza.

A discordant note was struck at the beginning of the constitutional talks. Lord Shepherd began by saying that a revised constitution for Gibraltar should make provision for the maximum control by the Gibraltarians over their domestic affairs within the constraints imposed by the following considerations: the constitution must not conflict, nor be seen to conflict, with the Treaty of Utrecht; it should not have the effect of worsening Anglo-Spanish relations; Britain must maintain an effective presence in Gibraltar's day-to-day administration if she was to adequately discharge her responsibilities, given her international and defence interests and Gibraltar's size and geographical location; there should be provision for the smooth reconciliation of British and local interests. It was on the second of these points, namely, the issue of Anglo-Spanish relations, that Lord Shepherd startled his hosts when he said that, 'We all hope that one day we may be able to achieve a satisfactory *modus vivendi* (with Spain over Gibraltar). Let us, therefore, do nothing that would make this impossible'. Sir Joshua Hassan replied to Lord Shepherd's opening remarks on behalf of the Gibraltar delegation and drew attention to the question of Anglo-Spanish relations as a constraint upon the impending constitutional debate:

> it was most unpalatable to be told at this crucial stage that we, who have been on the receiving end (of Spanish policies) all the time should not, by any new Constitution that we may work out together, make Anglo-Spanish relations any worse than they are.

He emphasised the need for formalising the link between Britain and Gibraltar:

> let there be no doubt that this aspect of the deliberations is uppermost in the minds of the people of Gibraltar today and in the days that will follow. It will be a sad day for Gibraltar if

at the end of these talks some acceptable formula has not been found.

Sir Joshua's anxieties were allayed during the course of the talks and the 'sad day for Gibraltar' that he had feared did not come to pass. When the discussions on a new constitution for the Colony ended they did so on a note of accord. An agreed final communiqué was released which covered both the link with Britain and the internal constitutional changes and detailed the nature of a legal instrument for the link, immigration into Britain, UK departmental responsibility for Gibraltarian affairs, integration with Britain and citizenship.

In the second part of the talks there was discussion of the constitutional proposals formulated by the Constitutional Committee of the Gibraltar Legislative Council and the Integration With Britain Party and the two delegations reached agreement on a number of changes to the existing Gibraltar Constitution. Significant among these changes, because of the increased autonomy which it bestowed, was the replacement of the existing Legislative and City Councils by a 'Gibraltar House of Assembly' with a life of four years and comprised of fifteen elected members and two ex-officio members, the Attorney-General and the Financial Secretary.

The twin objectives of the talks, the link between Britain and Gibraltar and a new constitution for the Colony, and the fact that the talks were conducted without recourse to the Spanish Government led Spain to denounce the deliberations in a statement issued to coincide with the final communiqué emanating from Gibraltar. According to the Spanish view, the constitutional talks had been 'seeking the adoption of a new constitution for Gibraltar in which the British military and colonial presence would be affirmed' despite the fact that the UN General Assembly resolution of 19 December 1967 had decreed that the 'colonial situation of Gibraltar could only be liquidated when Spanish national unity and territorial integrity were restored'. The statement stressed Spanish acceptance of the UN recommendation that the future of Gibraltar should be negotiated between Britain and Spain and pointed out that Britain had refused to comply and was even then seeking to give the Gibraltarians a sense of security through domestic legislation. Two scenarios were offered in the Spanish statement as to the outcome of the constitutional talks currently being held in Gibraltar.

Firstly, if the new measures that were proposed did not change Gibraltar's status they would serve to perpetuate a colonial situation condemned by the UN and Spain. Secondly, if the new measures did change Gibraltar's status they would unilaterally alter the Treaty of Utrecht and convert Gibraltar into Spanish territory under illegal and military occupation by a foreign power. In either case, the Spanish Government viewed any new agreements reached between Britain and Gibraltar on the Colony's constitutional status as a 'gratuitously unfriendly act' towards Spain, as a defiance of the UN and as a further obstacle to the attainment of a solution to Gibraltar's future.

The following day the Spanish statement was communicated to the UN Secretary-General, U Thant. With the UN again involved in the question of Gibraltar at the behest of Spain it fell to the British delegation at the Organisation to offer a reply to the Spanish charges. This was forthcoming in a letter to U Thant on 6 August. The British letter went to some lengths to point out that General Assembly resolutions did not constitute decisions binding on member states but only had the status of recommendations and that this voluntaristic element stood in contrast to the mandatory obligations which Britain had assumed under Article 73 of the UN Charter; to safeguard the interests of non-governing territories for which it was responsible. The British delegation emphasised that in a case of incompatibility between the Charter of the UN and a resolution of the Organisation's General Assembly it was the former obligation which had to take precedence.

On the specific point of the new constitutional arrangements for Gibraltar the British reply stated that these took due regard of the particular circumstances of the Gibraltarians and the freely and democratically expressed views of their representatives; that the new arrangements did not alter the international status of Gibraltar, nor create further obstacles to an eventual solution acceptable to its people, Britain and Spain. The British Government was said to be ready to engage in talks with Spain as soon as there was real evidence that such talks were likely to be constructive but that while Spain continued to exert 'crude pressure', disregarded the 'distrust and dislike' which that pressure created and failed to appreciate the need for 'patient endeavour' such a prospect seemed remote. In the meantime the expressed intention of the British Government was to continue to safeguard the interests of the Gibraltarians.

The UN was to remain the central focus as the year ended. On 18

December the General Assembly adopted a resolution on Gibraltar by 67 votes to 18 with 34 abstentions. The resolution declared that: the 'colonial situation in Gibraltar (was) incompatible with the UN Charter and previous General Assembly resolutions'; called upon the UK to 'terminate the colonial situation in Gibraltar' not later than 1 October 1969; and called upon the British and Spanish Governments to begin negotiations to this end 'without delay'.[8]

Clearly, the Spanish Government's arguments had again won the day at the UN. Equally apparent was the failure of Britain to engage in successful coalition building in support of its case. Still more troubling was the second part of the General Assembly's resolution which called upon the UK to 'terminate the colonial situation in Gibraltar', for not only was this an explicit instruction, which carried the implication that Britain should cede Gibraltar to Spain in accordance with the Treaty of Utrecht, but also emphasised that this be carried out within a specified time period; i.e. by 1 October 1969.

Hardly surprisingly, Lord Caradon, told the General Assembly that the resolution it had approved: 'will not and cannot be put into effect' before adding that any attempt to force Britain to hand over the inhabitants of Gibraltar to Spain against their will was 'happily so removed from possibility as to be incredible'.

5

CONTINUING GLOOM

Despite the UN General Assembly's resolution on Gibraltar on 18 December 1968, the British Government pressed ahead with the constitutional changes agreed earlier that year and, on 30 May 1969, the new constitution for Gibraltar was published in London and in the Colony as an Order in Council with immediate effect.

The maintenance of Gibraltar's link with Britain was formalised in the preamble to the new constitution by reaffirming that Gibraltar was 'part of Her Majesty's dominions'; that HMG had 'given assurances to the people of Gibraltar that Gibraltar would remain part of Her Majesty's dominions unless and until an Act of Parliament, provides otherwise'; and that HMG 'will never enter into arrangements under which the people of Gibraltar would pass under the sovereignty of another State against their freely and democratically expressed wishes'.[1]

Such assurances demonstrated that the British Government continued to feel itself unrestricted in its dealings with Gibraltar by the General Assembly resolution and still viewed that resolution as a recommendation with which it felt unable to comply rather than as an instruction which was binding upon it. Predictably, even inevitably, the Spanish Government adopted a markedly different stance. The new constitution for Gibraltar was described by the Foreign Minister, Señor Castiella, as reflecting 'open disregard' by Britain of the UN resolution and as a violation of the Treaty of Utrecht. The Spanish Cabinet met on 6 June and what it termed the 'so-called' constitution of Gibraltar was the major item on its agenda. The outcome of the Cabinet's deliberations became evident two days later when the Spanish authorities closed the customs post and frontier crossing to Gibraltar at La Linea; an action described in an official statement released in Madrid as having been taken 'to defend Spanish interests and rights in Gibraltar'. It is difficult, if not impossible, to discern a rational link here between the end sought and the means employed.

The closure of the frontier produced immediate consequences for the Spanish workers who had daily crossed into Gibraltar from Spain. Unable now to go to work, the 4,800 workers were promised by the Spanish Government that they would receive equivalent wages or social security payments in Spain.[2] No less immediate was the consequence of the frontier closure for Gibraltar. The fact that so many could no longer go to work meant that the Colony was instantly deprived of roughly one-third of its workforce. Contingency plans, which had been prepared since the first restrictions had been imposed at the frontier in 1964 and which were intended to cover this very eventuality, were now put into effect. The Gibraltar Chamber of Commerce opened a centre to assist its members in finding replacements for the Spanish workers whose services they had lost; British troops based in Gibraltar were called in to help maintain essential services; and employees of the Gibraltar Government were actively encouraged to take on additional part-time jobs.

To consolidate these initial measures, on 10 June, the Gibraltar Legislative Council passed an ordinance under which persons working in essential services were not permitted to change their employment without the approval of the Director of Labour and Social Services. Similarly, employers were precluded from engaging workers already employed in essential services without special permission having been sought and granted by the Director.[3] These measures enabled *The Times* correspondent in Gibraltar to report that everything in the Colony was 'absolutely normal', except for the closure of a few Spanish-owned shops, that stevedores were arriving from Morocco to keep the port's cargo-handling manpower at full complement and that local women were filling the vacancies in shops and restaurants created by the forced withdrawal of Spanish employees.[4]

The Foreign Secretary, Mr Michael Stewart, told the House of Commons, on 9 June, that the Governor of Gibraltar, Admiral Sir Varyl Begg, was confident that essential services would be maintained, although temporary difficulties could be anticipated as a result of a loss of 30 per cent of the Colony's labour force. Mr Stewart emphasised that, 'As the Governor has said, Gibraltar's watchword is "business as usual" . . . It remains the declared policy of HMG to support Gibraltar, and we will continue to sustain its people.' Perhaps as a sop to public morale in Gibraltar, Sir Joshua Hassan announced, on 20 June, that the first general election to be held under the new constitution would take place on 30 July.

With the frontier closed and Gibraltar deprived of one-third of its workforce the Spanish Government took the decision to tighten the cord which it had drawn into place around the Rock. On 25 June the British Ambassador in Madrid was informed by the Spanish Foreign Ministry that it was suspending the ferry service between Algeciras and Gibraltar from 27 June.[5] The justification for this action was stated to be 'the defence of Spanish interests'.

In a statement to the House of Commons on 26 June, the British Foreign Secretary, Mr Stewart, described the Spanish action in closing the Algericas–Gibraltar ferry as one which 'flouts the standards of international behaviour accepted by modern governments, and will do nothing whatever to bring a solution to the Gibraltar dispute nearer'. Mr Stewart also revealed that the latest Spanish action had been foreseen, 'there was reason to believe that the Spanish Government would shortly withdraw the ferry operating from Algeciras'. Alerted to that possibility he had instructed the British Ambassador in Madrid to deliver a Note to the Spanish Government on 24 June 'requesting agreement in principle for the operation of a Gibraltar-run ferry service between Gibraltar and Algeciras'. It was now the turn of Britain to cite the terms of the Treaty of Utrecht, something which up until then had been largely the preserve of the Spanish Government, and Mr Stewart told the House that: 'The interruption of normal maritime communications between Gibraltar and Spain is . . . contrary to the Treaty of Utrecht, in which it is implicit that there should be sea communication between Gibraltar and Spain'.

On 27 June, Mr Stewart summoned the Spanish Ambassador in London, Marques de Santa Cruz, to the Foreign Office and handed him an *aide-mémoire* which pointed out 'the futility of the apparent intention of the Spanish Government to reduce Gibraltar by malicious pressure exerted on the population'. The *aide-mémoire* was dismissive of Spanish claims to the sovereignty of Gibraltar and highlighted a credibility gap which the British Government discerned as existing between the Spanish claims made over Gibraltar at the UN and the reality of Spanish actions towards Gibraltar:

the Spanish Government cannot reasonably believe that any

British Government could hand over the people of Gibraltar to a Government which has done them so much harm already and which demonstrates unrelentingly that its immediate object is to disrupt the daily life of Gibraltar and destroy the people's livelihood. The role which the Spanish Government attempts to play at the United Nations as the willing guarantor of the interests of the people of Gibraltar is simply not credible.

In response to the challenge to its credibility on 4 July the Spanish Foreign Minister, Dr Manuel Fraga Iribarne, announced, that as a result of a Cabinet meeting, Spain would publish a decree offering Spanish nationality to all residents of Gibraltar and persons born there.[6] *The Times* correspondent in Gibraltar described the reaction of the Rock's inhabitants to the offer contained in the Spanish decree as having:

> given the Rock a big laugh – that and a few unprintable words ... After five years of restrictions and hardships at the hands of the Franco regime in trying to bring the people of the Rock to their knees, which has only resulted in making them more British than the British, this offer is regarded here as completely and utterly ludicrous.[7]

The continuing and growing pressure exerted by Spanish frontier restrictions upon Gibraltar's economy, and with no immediate prospect of an early alleviation of the Rock's difficulties, may have been significant factors in bringing about a change in the Colony's political leadership. On 30 July the first general election under the new constitution was held for the fifteen member House of Assembly and produced an inconclusive result with no party obtaining an overall majority. Sir Joshua Hassan's Gibraltar Labour Party secured seven seats; one short of an overall majority.[8] The Integration With Britain Party of Major Robert Peliza won five seats and the three remaining seats went to a group of Independents led by Mr Peter Isola, a former Deputy Chief Minister. The Governor called upon Sir Joshua to form a Government but talks aimed at producing a coalition between the Gibraltar Labour Party and Mr Isola's Independents broke down. On 16 August, Major Peliza was able to inform the Governor that his Integration With Britain Party and Mr Isola's Independents had reached agreement on the form of a coalition Government and, accordingly, Major Peliza succeeded Sir Joshua as Chief Minister. The strength of the

Integration With Britain Party's showing in the election was seen by some to stem from the fact that many Gibraltarians credited Major Peliza with being the politician who had won the assurances about the Rock's future links with Britain contained in the preamble to the new constitution.

This element of instability in the politics of Gibraltar introduced an unknown factor into the relationship with Britain which, to many outside observers, had owed its strength to the continuity of leadership in Gibraltar which had been provided by the repeated electoral successes of Sir Joshua's party and his own enduring hold upon the city's key political position. Now a new political leader in Gibraltar would have to negotiate with London and London would have to seek to accommodate an even more overtly pro-British posture emanating from Gibraltar. However, events were to dictate that growing Spanish pressure on Gibraltar would serve to override any significant re-orientation of Anglo-Gibraltarian relations which the electoral outcome might otherwise have heralded.

The new Gibraltar Constitution, under which the general election in the city had been conducted, continued to provide a point of contention between Spain and Britain. On 16 June, the Spanish permanent representative to the UN, Señor de Pinies, sent a Note to the UN Secretary-General, U Thant, which contended that the new Gibraltar Constitution involved 'not only a disregard for the recommendations of the General Assembly on the way to put an end to the colonial situation in Gibraltar, but a strengthening of this situation through the creation of an artificial obstacle in the way of [those recommendations]'. It was also claimed that the new constitution violated Article X of the Treaty of Utrecht. U Thant was the recipient of a counter British Note on 5 August which described Spain's closure of the land frontier with Gibraltar as an 'attempt . . . to impose a form of economic siege' which 'belies the concern which the Spanish Government professes for the interests of the people of Gibraltar'. The British Note denied that Gibraltar's new constitution effected any change in the city's international status nor that it entailed any breach of the Treaty of Utrecht. A further Spanish Note to U Thant, on 4 September, alleged that the British Government 'hesitates to enter into sincere and meaningful negotiations on the future of Gibraltar because its only interest is in perpetuating its military presence at a nerve centre in the geography of Spain'.

The 'form of economic siege' to which Gibraltar was subjected by the frontier restrictions served to increase the city's dependence upon aid received from Britain. Lord Shepherd, Minister of State for Foreign and Commonwealth Affairs, held talks in Gibraltar, on 5–8 September, about the Rock's economy with the Governor and Major Peliza, the recently appointed Chief Minister. As a result of those talks the Foreign and Commonwealth Secretary, Mr Stewart, was able to tell the House of Commons that the Gibraltar authorities were revising the Rock's development plan and that further talks would take place before the end of the year.[9]

Perhaps out of frustration that the battle of words over Gibraltar had proved to be fruitless with each claim being seemingly matched and nullified by a counter-claim the Spanish and British Governments deployed large naval forces off Gibraltar at the end of September and the beginning of October as though to each test the resolve of the other by some more tangible means than the diplomatic Note or *aide-mémoire*.[10]

Despite this display of naval might, of greater immediate importance to the people of Gibraltar, and of greater impact upon them, was the Spanish Government's decision to cut the telephone links between the Spanish mainland and the city on 1 October – the deadline given in the UN General Assembly's resolution of 18 December 1968, as the date by which Britain was to 'terminate the colonial situation in Gibraltar'. Although no official announcement was made in Madrid of this latest action, the British Embassy in the capital had been made aware of it by the main telephone exchange and, in anticipation of it, the Gibraltar authorities had established alternative telephone links through Malta and Morocco in case the two direct lines to London, passing through Spain, were also cut. In London itself a spokesman for the Foreign and Commonweath Office described this latest Spanish action as 'another petty, malicious and small-minded restriction in complete contrast with the British Government's policy, which has never been to break the link between the Gibraltarians and the people of Spain', and the Spanish Embassy was told of the British Government's disapproval of 'this malicious action . . . to intimidate a small community'.

The Spanish action was defended by Foreign Minister Dr Castiella, in a Note to the UN Secretary-General of 1 October in which it was declared that Spain would 'defend, with whatever means it deems to be appropriate, its inalienable right to the unity and integrity of its national territory'. Dr Castiella accused the British Government of having created a situation in the Gibraltar area 'whereby an incident might occur at any time because of its constant manoeuvres and military activity'.[11] In response to these Spanish claims, Lord Caradon, UK permanent representative to the UN, wrote to U Thant and denied that Britain retained sovereignty over Gibraltar for imperial or military reasons and charged the Spanish Government with deliberately seeking to intimidate the people of Gibraltar by various means of 'harassment and pressure'.[12]

Fortunately, the naval show of strength which was even then underway off Gibraltar did not lead to any incidents, although it did serve to introduce a level of antagonism into Anglo-Spanish relations over Gibraltar which had not hitherto been evidenced and suggested a qualitative escalation in the stakes which each nation was prepared to wager in pursuit of their respective claims to sovereignty of the Rock.

From necessity, but possibly also to make Spanish 'harassment and pressure' more tolerable and bearable to Gibraltar, the British Government announced a three-year, £4 million grant of development aid for the city at the end of a visit to Britain by Major Peliza in December.[13] Accompanying the announcement was a reiteration of the assurances given in the preamble to Gibraltar's new constitution that the Rock remained part of the Queen's dominions and that no change of status would be effected against the wishes of the Gibraltarians themselves.

Events of the last two months of 1969 stood in marked contrast to the militaristic posturing which had been a feature of late September and the early days of October. On 28 October there had been an extensive reorganisation of the Spanish Cabinet as a result of which Foreign Minister Dr Castiella, the man who had masterminded Spain's 'hard-line' policy on Gibraltar, was succeeded by Señor Gregorio Lopez Bravo. A shift in the orientation of Spanish policy towards Gibraltar was quickly evidenced when, immediately upon Señor Bravo assuming office, the broadcasting of propaganda to Gibraltar from a nearby Spanish radio station ceased. Eight days later, Britain's newly appointed ambassador to Spain, Sir John Russell, had what he described as an 'extremely cordial' meeting with the Spanish head of state, General Franco, during which it emerged that Spain wished to adopt 'a more pragmatic and practical approach' to its differences with Britain in the hope that relations between the two countries could flourish in a 'warmer climate'.[14] These two events clearly indicated that the Gibraltar question was to be relegated to a lower position on the Anglo-Spanish agenda than that which it had occupied since 1964; a view confirmed on 18 November when the Spanish Ambassador to the UN announced that his Government did not intend to raise the question of Gibraltar during the then current session of the General Assembly.

Further confirmation of Spanish wishes to improve relations with Britain and, therefore, the necessity of moderating the previous stance adopted on Gibraltar was forthcoming on 19 December when Señor Bravo said, 'I firmly desire a relaxation of tension with the UK'. He added that although the Spanish claim to Gibraltar was still an important element in his country's foreign policy it was 'not the magnetic north that the pointer always seeks'.[15] However, Señor Bravo was more guarded on the subject of the frontier restrictions imposed upon Gibraltar by the Spanish authorities. He insisted that these did not constitute a blockade but simply amounted to a partial re-establishment of the 'absolute cutting-off of communications between Gibraltar and the rest of the peninsula which was agreed upon in the Treaty of Utrecht'. A thaw in the icy climate of Anglo-Spanish relations was not about to lead to a flood.

Nevertheless, the impasse which had existed between London and Madrid showed every sign of being vulnerable to the growing spirit of cordiality which became a feature of the following year as rumours circulated of behind-the-scenes talks to reach a solution to the Gibraltar problem.[16] However, on 13 May the Spanish Government issued a statement denying these rumours[17] and although Lord Shepherd expressed the desire of the British Government to move to a solution of the Gibraltar problem he insisted that an essential pre-condition for any such solution was the lifting by Spain of the frontier restrictions. Despite talks between Sir Alec Douglas-Home, Foreign Secretary in the recently formed Conservative Government, and Señor Lopez Bravo in the middle of the year at the British Embassy in Luxemburg and between Sir Alec and Major Peliza in London in the summer, no positive signs were to be seen to develop as a result of the much-vaunted cordiality which had entered Anglo-Spanish relations in the previous twelve months.

In Gibraltar, where the full impact of Spanish frontier restrictions continued to be felt, small and fragile intimations of optimism could be discerned in 1970 which were far less apparent to the policy makers in London and Madrid. No doubt, such optimism was partly born of despair, for in August, Mr Lewis Stagnetto, President of the Gibraltar Chamber of Commerce, stated that, 'the off-the-cuff policies of the present Government are leading us down the path to ruin and, unless a stop is put to this downward course in the very near future, the end of Gibraltar is in sight'. Mr Stagnetto was, however, careful not to lay blame for Gibraltar's parlous economic situation exclusively at the door of the city's present Government. The first sign for optimism had been detected in Gibraltar during a visit made to the adjoining Spanish region of Algeciras by the Spanish Minister of Industry, Señor Jose Maria Lopez de Letona, to foster industrial development, for it was apparent that this area of Spain was itself suffering economic hardship as a consequence of its trading links with Gibraltar having been cut.[18]

The second cause for optimism was seen in the city as arising from the negotiations then being conducted in Brussels to take Britain into the EC. On 1 October 1970 Spain had become linked to the Community through a preferential trade agreement and any new accord which linked Britain and Gibraltar to the Community was viewed as obliging Spain to open its frontier with the Rock to allow the free passage of goods.[19]

Such optimism began to look somewhat ill-founded in February 1971 when the contents of a speech by the Spanish Foreign Minister to the Advanced College for National Defence Studies in Madrid were released. Señor Bravo described Gibraltar as presenting a threat to the security of Spain, both as a military objective which could cause fighting on Spanish territory and as a base from which Spain could be attacked. He also repeated the Spanish claim to sovereignty over Gibraltar but did not rule out the prospect of further Anglo-Spanish negotiations: 'We are not hostile to Great Britain; it is both desirable and possible that the two countries should reach a final solution to their old quarrel over Gibraltar at the negotiating table in the interests of European unity.'[20]

Nor was there any further cause for optimism over a solution to the Gibraltar question in June when the informal talks held in Madrid between Sir Denis Greenwood, Permanent Under-Secretary at the Foreign and Commonwealth Office, and Señor Bravo ended with the two sides' attitudes remaining unchanged.[21] No more hope for a swift end to Gibraltar's hardships emerged from a visit made to the city by the British Foreign Secretary for talks with Gibraltarian Government and Opposition leaders.

In October the UN provided the venue for direct talks between the British and Spanish Foreign Ministers[22] after which Señor Bravo was reported as believing that an 'atmosphere for dialogue' now existed between Madrid and London and that negotiations on Gibraltar were possible.[23] However, as Señor Bravo continued to insist that 'the question of sovereignty over the Rock, which we cannot renounce, is a matter that can only be discussed between the Governments of Spain and Great Britain, and not by the Gibraltarians' and while Britain continued to honour the assurances given in the preamble to the new Gibraltar Constitution to do nothing against the wishes of the city's inhabitants, an insurmountable barrier still seemed to stand in the way of a solution to the Gibraltar question.

Indeed such proved to be the case in February 1972 when unproductive talks were held in Madrid between the two Foreign Ministers. On his arrival back in Britain, Sir Alec conceded that the problem of Spanish claims on Gibraltar was still 'intractable', that he had not expected any tangible results from his meeting in Madrid and that there had been 'no promises' on any removal of Spanish sanctions. He added that:

> The situation [in Gibraltar] now is very unsatisfactory. There ought to be contact with the Spanish mainland. I think the Spaniards genuinely want a new deal with Britain which would include arrangements for Gibraltar. We shall look at any proposals they make to see if they would suit the people of Gibraltar – to see if there's any change of status.[24]

During Sir Alec's visit to Madrid a further round of 'explanatory talks' was decided upon and set to be held between the two Foreign Ministers in London in July 1972. In the interim, and a year earlier than had been scheduled, Gibraltar went to the polls in what was to be the most acrimonious election in the Rock's history.

The election had been occasioned by the dissolution of the House of Assembly on 22 May at the request of the Chief Minister. Ostensibly the Chief Minister had requested the dissolution because of his lack of confidence in Major Alfred Gache, Minister for Commercial Economic Development, and one of the three independents with whom the Integration With Britain Party had formed a coalition Government after the last general election of July 1969. Major Gache resigned from the Government on 1 June and did not contest the election held on 23 June. However, behind the ostensible reason given for the request for the dissolution of the House of Assembly, the Gibraltar press discerned another less manifest rationale. According to press speculation the real reason underlying Major Peliza's decision to call an early election was the suspicion that the Leader of the Opposition and long-serving former Chief Minister, Sir Joshua Hassan, was prepared to accept a rumoured Spanish offer to grant Britain a long lease on Gibraltar in exchange for recognition of Spanish sovereignty over the Rock.[25]

Inevitably, given the forthcoming talks to be held in London in July between the British Foreign Secretary and the Spanish Foreign Minister, the future of Gibraltar became the major issue of the election campaign with the Integration With Britain Party, supported by the Transport and General Workers Union[26] accusing the Gibraltar Labour Party of a willingness to accept a compromise settlement in the Anglo-Spanish dispute and the latter issuing strong denials of this allegation and rejecting any suggestion of a preparedness to surrender on the question of sovereignty.

In the event Major Peliza's gambit of calling an early election failed and the Gibraltar Labour Party won a clear majority in the House of Assembly. On 25 June, the Governor of Gibraltar appointed Sir Joshua Hassan as the new Chief Minister and the Rock's experience of coalition Government ended. Hardly surprisingly, Sir Joshua claimed that the electoral outcome was a vote of confidence in his party and in his leadership after the 'unfortunate' campaign which had been waged against him personally by the Integration With Britain Party. He denied any 'conspiracy' between Britain and Spain, with himself acting as a 'fifth column', to arrange a settlement prejudicial to the Rock's association with Britain and British sovereignty.[27]

The British and Spanish Foreign Ministers met in London on 9–21 July as agreed at their last meeting in Madrid in the previous February. Gibraltar was discussed 'in a friendly and constructive spirit' and Sir Alec Douglas-Home agreed to study Señor Lopez Bravo's summation of Spanish perceptions of the Gibraltar question with a view to taking the matter further at a 'working meeting' later in the year. In August the Governor of Gibraltar and the recently elected Chief Minister visited London to be appraised of the latest developments in the conversations between Britain and Spain over the Rock by Sir Alec Douglas-Home and were reassured that Britain stood by its commitment to consult the Gibraltarians before proposing any changes to the status of Gibraltar. The 'working meeting' between the two Foreign Ministers took place in Madrid on 27–28 November and a joint communiqué issued after the talks said that they had 'not yet reached the stage at which formal negotiations might begin'.

The year 1973 saw no departure from the pattern established over the course of the previous two years. In January Mr Julian Amery, Minister of State at the Foreign and Commonwealth Office, paid a three-day visit to Gibraltar during which he claimed that the 'siege' which Spain had imposed upon the City for the last three and a half years 'to coerce Gibraltar into submission' had been a complete failure. The Spanish Foreign Minister, Señor Lopez Bravo, conceded that the dispute over Gibraltar was seriously damaging Anglo-Spanish relations and that the Spanish people were tiring of the unproductive dialogue, although he personally was still optimistic about his series of talks with his British counterpart and said that he would be tabling some new proposals at their next meeting.[28] That meeting took place in London on 8–9 May and ended in deadlock because the British were unable to accept the Spanish proposals as a basis for discussion. No official communiqué emanated from the meeting but Señor Lopez Bravo emphasised that the Anglo-Spanish dialogue had merely been postponed and not terminated.

Whether 'postponed' was a more accurate description than 'terminated' to describe the then current state of Anglo-Spanish discussions on the problem posed by Gibraltar is perhaps a moot point given that the problem seemed manifestly intractable. All possible solutions to the problem appeared to have been considered in the numerous rounds of talks which had taken place, and every solution offered had been found to be unacceptable to one or other of the two parties. It now appeared that the way ahead was to be found in some development beyond the strict confines of Anglo-Spanish relations and could hinge upon Spain entering NATO, which seemed improbable, or upon Spain joining Britain in the EC, which seemed to lie a long way in the future. For the present, an impasse had been reached and this was attested to by the replacement of the anglophile and European-oriented Señor Lopez Bravo as Spanish Foreign Minister by Señor Laureano Lopez Rodo shortly after the former's return from the last round of Anglo-Spanish talks held in London.[29]

This change of personnel at the top-most level of the Spanish Foreign Ministry heralded a return to the 'hard-line' policy which had characterised Señor Castiella's period as Foreign Minister for, within five weeks of his appointment, Señor Rodo informed the UN Secretary-General, Dr Kurt Waldheim, that discussions with Britain over Gibraltar had broken down because 'more than four and a half years have passed without the British Government having made any effort to fulfil the UN resolution [on Gibraltar] of December 18,

1968'. Señor Rodo alleged that the British Government had 'shielded itself behind the preamble of a constitution which it had dictated and imposed after the UN resolution with the deliberate intention of making decolonisation more difficult' whilst Spain had put forward a plan 'to safeguard the interests of the Gibraltarian population when the colonial situation ended in conformity with UN resolutions'. The Spanish Government's view was that it now had no option but 'to give very serious thought to what further steps it should take in connection with this problem'.[30]

These 'further steps' were soon decided upon and taken by the Spanish Government. On 6 August, British yachts and other vessels registered in Gibraltar were forbidden to enter the port of Algeciras; this had become the normal means of communication between the Rock and the Iberian peninsula. On 26 November, Spain again went to the UN with a renewed call to the Trusteeship Committee for a British withdrawal from Gibraltar whilst again rejecting the British offer to refer the question of Gibraltar to the International Court of Justice because the problem was of a political and not of a legal nature.

Again, on 14 December, the UN General Assembly had adopted a consensus, on the recommendation of its Trusteeship Committee, which repeated 'the hope that negotiations with a view to the final solution to [the Gibraltar] problem . . . will soon be resumed by the United Kingdom and Spain' and had called upon both countries 'to spare no effort in order to arrive at a solution consonant with the principles of the (UN) Charter and to report back on the result of these negotiations to the Secretary-General and to the General Assembly at its 29th session'.

The year ended with a marked absence of the cordiality and dialogue which had characterised Señor Lopez Bravo's occupancy of the office of Spanish Foreign Minister and yet again the UN had been called into the dispute.

6

A GLIMMER OF HOPE

In accordance with the most recent General Assembly consensus on Gibraltar, British and Spanish representatives met in Madrid on 30–31 May 1974 for talks which were described by the Foreign Office in London as 'exploratory and non-committal'. The Madrid meeting had been proposed by the British Government on 11 April but few could have entertained serious hopes that a positive outcome would emerge as there was nothing to indicate that the attitudes of either party had changed since the breakdown of the last round of talks in 1973.

The leader of the British delegation, Sir John Killick, Deputy Under-Secretary of State for European Affairs, described the talks as 'useful' but how useful they really had been can be gauged by the fact that no decisions were taken nor any date fixed for a subsequent meeting. Señor Juan Jose Rovira, Under-Secretary at the Spanish Foreign Ministry and leader of his country's delegation, was more forthcoming on the failure of the talks to make progress. On 31 May he told reporters that the British delegation had been especially concerned to raise the matter of access to Gibraltar airport but that the Spanish delegation had not felt itself able to respond on this specific issue because 'it was not prepared to accord facilities which would serve only to consolidate the British colonial presence on Spanish territory'. The Spanish view was that agreement on the use of Gibraltar airport could only be reached as part of an overall settlement on the Rock's decolonisation.[1]

With bilateral talks between Britain and Spain on Gibraltar's future again deadlocked, the UN provided the setting for the next exchanges on the intractable problem that the Rock posed. On 3 October, Spain's Foreign Minister, Señor Pedro Cortina

Mauri, restated his country's position on Gibraltar to the General Assembly. He called upon the General Assembly to remind Britain of its 'obligation to negotiate with Spain the decolonisation of Gibraltar by reintegrating this territory into the Spanish nation' and added that the British delegation at the talks in Madrid in May had been unwilling to deal with matters of substance. Britain's military base in Gibraltar was said by Señor Mauri to be a threat to Spain and he accused the British Government of being 'obstinate, rigid and selfish' in its refusal to negotiate. He was adamant that the continued occupation of Gibraltar was 'a violation of the territorial integrity of Spain, a danger to its security, interference in its external policy of peace, and a barrier to bringing up-to-date the legal regulation of the waters of the Strait [of Gibraltar]'.

The British permanent representative at the UN, Sir Ivor Richards, told the General Assembly that the British Government had 'no intention whatsoever of handing over the people of Gibraltar, against their own wishes . . . to a country which a generation ago turned its back on democracy and the democratic process'. Mr Richards emphasised that the Spanish Government had been 'singularly – one could almost say spectacularly – unsuccessful in persuading the people of Gibraltar that it is in their interests to give up their present status'. He did, however, state that Britain was ready to renew talks with Spain over Gibraltar.

This exchange, with the words of both speakers containing more than a little venom, marked an abrupt termination in Anglo-Spanish negotiations over Gibraltar. Bilateral talks on the future of the Rock were not to be resumed until the autumn of 1977 by when a new political regime had made its appearance in Spain.

Although face-to-face talks at government level between Britain and Spain were now in abeyance the debate on Gibraltar's future continued through the unlikely medium of the letters' page of the *The Times* newspaper. On 30 September, the newspaper published a two-page special report which provided the background to the long-standing Anglo-Spanish dispute over the Rock in which 'the views of the principal parties involved' were presented. The Spanish position was outlined in an article by Señor Juan Roldan in which he emphasised the permanence of the claim to the sovereignty of Gibraltar in Spanish foreign policy; raised strong doubts as to the right of Gibraltarians to decide the question of sovereignty 'since they form the civilian population of a foreign military base' and 'strictly speaking, do not constitute a national entity'; made mention of the 'special regime of legislative, judicial, administrative and financial autonomy' which Gibraltarians would enjoy under Spanish rule; and concluded that:

> the intransigent attitude of successive British Governments towards restoring Gibraltar to Spain does not appear to be very reasonable, unless it be that the reference to the wishes of the people of Gibraltar is simply an excuse for consolidating a military base which controls one of the entrances to the Mediterranean.[2]

On 14 October *The Times* published a letter from Sir Joshua Hassan, in which he stated 'the essential points of the Gibraltar case' which he believed had been inadequately delineated in the report of 30 September. Sir Joshua rejected Señor Roldan's description of the Gibraltarian population as having been 'artificially imported' by the British and stressed instead the 'distinct entity' of the Rock's community. He highlighted Britain's renewal at the UN General Assembly of its pledge to abide by the wishes of the people of Gibraltar in any negotiations with Spain. Sir Joshua emphasised that:

> the Spanish Government denies that we have any right at all to express our wishes and claims to protect our interests – but the Spanish Government is to be the arbiter of what our interests are. Much is made of a special regime for the Gibraltarians but no details are disclosed.[3]

Dr Manuel Fraga Iribarne, the Spanish Ambassador in London, entered the debate with a letter to *The Times* of 23 October in which he argued that there was historical evidence to support the view that the inhabitants of Gibraltar constituted an 'artificial' population whilst dismissing Sir Joshua's reference to Spain's disregard for the right of the Gibraltarians to express their wishes as 'quite inaccurate' since 'Spain simply denies that [the Gibraltarians] have any right to make decisions with regard to a part of Spanish territory that has never belonged to them, and to which at no time have they held any legal title'. But it was in directly challenging Sir Joshua's allegation that there had been no disclosure of the details of the special regime proposed by Spain for Gibraltar that the real thrust of Dr Manuel Fraga Iribarne's letter lay:

This supposed ignorance could be remedied through the authorities responsible for Gibraltar's external affairs. In any case, full and detailed information concerning a statute of this kind was given, at the beginning of 1973, to a distinguished Gibraltarian who is a prominent figure in the public life of the Rock. That he did not think it proper to make it known to his fellow citizens is hardly Spain's fault.[4]

Dr Manuel Fraga Iribarne's assertion that 'a distinguished Gibraltarian who is a prominent figure in the public life of the Rock' had been the recipient of 'full and detailed information relating to Spanish proposals for Gibraltar's future', which had not been revealed to the people of the city, threatened to renew allegations which had featured prominently in the campaign leading up to the general election in Gibraltar of 23 June 1972. It will be remembered that the then Chief Minister of Gibraltar and leader of the Integration With Britain Party, Major Robert Peliza, had contended that Sir Joshua, at that time leader of the Opposition on the Rock, had been involved in a conspiracy between Britain and Spain to arrange a settlement prejudicial to Gibraltar's association with Britain and continued British sovereignty.[5]

It would seem inconceivable that such Spanish proposals would have been put without there first having been fairly lengthy and detailed discussions held between the Spanish authorities and one or more Gibraltarian representatives and, in all likelihood, representatives of the British Government. Dr Manuel Fraga Iribarne's revelation now seemed to cast doubt upon the integrity of the present Government of Gibraltar and specifically questioned the character of its Chief Minister.

The key points in the Spanish list of proposals for a new regime in Gibraltar were dated 28 February 1973 and were summarised in *The Times*.[6] They stipulated that:

1 As soon as Spanish sovereignty of Gibraltar was recognised then the neighbourhood of Gibraltar would become a special territory with legislative, judicial autonomy – administrative and financial.
2 Gibraltarians would take Spanish nationality but would not need to renounce their British nationality except if this were required by British law.
3 The Spanish legal system 'as developed by the special legislation of Gibraltar' would apply after the 1969 Gibraltar Constitution

had been suitably amended and that the Spanish penal and police laws would be introduced in all areas concerning Spain's internal and external security.

4 The senior authority in Gibraltar would be a civil governor appointed by the Spanish head of state.

5 The most senior members of the executive would be Spaniards or Gibraltarians of Spanish nationality.

6 Spanish would be the official language of Gibraltar whilst the use of English would be safeguarded.

Details of the Spanish proposals had been made available to *The Times* by Sir Joshua Hassan and in the same edition of the newspaper there appeared a further letter from him in which he admitted that he had been handed the Spanish proposals by a representative of the Spanish Foreign Ministry in February, 1973, whilst on a visit to Brussels to discuss EC business and that his meeting with the Spanish representative 'took place with the knowledge of the then [British] Secretary of State'.[7] Sir Joshua contended that there was 'nothing new in contacts between representatives of Gibraltar and of the Spanish Government' but that on reading the Spanish proposals he had informed the Spanish representative 'that there was no question of the regime described to me being acceptable to the people of Gibraltar', especially in the light of the 1967 referendum in which the Gibraltarian electorate had overwhelmingly rejected any transfer of sovereignty to Spain. Sir Joshua said that he had not felt able to reveal the Spanish proposals because he felt that his meeting in Brussels had been 'governed by the diplomatic convention that meetings held in confidence are regarded as confidential' but that the situation had now changed because 'the ambassador's letter clearly absolves me from any such restriction'.

Two of the admissions contained in Sir Joshua's letter of 7 November are worthy of note namely, his admission that there was nothing new in contacts between Gibraltarian representatives and the Spanish Government, and his further admission that he had not publicly revealed the content of the Spanish proposals passed to him in Brussels. Therefore, it was hardly surprising that both of these points were taken up by the Leader of the Opposition in Gibraltar, the Leader of the Integration With Britain Party, Mr Maurice Xiberras, in a letter to *The Times*.[8] Mr Xiberras pointed out the efforts made by his Party 'in and out of the Gibraltar House of Assembly' to discover details of the Spanish proposals and the circumstances of the Brussels meeting and stressed that:

> contacts between Gibraltar ministers, ex-ministers and prominent Gibraltarians with Spanish Government representatives and the 'proposals', 'counter-proposals', 'new ideas' and 'certain ideas' which it was suspected were being discussed, have been a fundamental issue in Gibraltar since before the general election of 1972.

Mr Xiberras also claimed that the convention which dictated that he, as Leader of the Opposition, should have been consulted on all discussions with the Spanish Government, had been flouted. This fact had led him to protest to both the Governor of Gibraltar and the British Foreign Secretary in the autumn of 1972 that he had been excluded from learning of the content of discussions emanating from the visit of the Spanish Foreign Minister, Señor Lopez Bravo, to London, earlier that year.[9] Two further areas of concern were mentioned by Mr Xiberras. Firstly, 'what else transpired at this meeting in Brussels, which has been kept a secret for twenty months, which might conceivably affect the interests of our people?' Secondly, 'whether other meetings have taken place and whether there have been any modifications of those proposals of the Spanish Government through the same or other channels?'[10]

Neither of these concerns elicited a response from the Governments of Gibraltar, Spain or Britain; although in Madrid it was held that the Spanish proposals were more favourable to Gibraltarians than Sir Joshua had stipulated since it was pointed out that the legal framework to be established in a Spanish-ruled Gibraltar would take into account existing international treaties including the treaty that would have to be agreed between Spain and Britain before any transfer of the sovereignty of the Rock could take place.

A number of reasons may be suggested as to why the Spanish Government's proposals on Gibraltar were made public by Sir Joshua Hassan when they were. Firstly, it could have been that the pressure exerted by the Opposition in the Gibraltar House of Assembly, the Integration With Britain Party, had forced Sir Joshua to make his revelation; certainly Mr Xiberras mentioned such pressure in his letter to *The Times*.[11] This, however, leaves unanswered the question 'why at this time did the Spanish proposals enter the public domain?' since, by his own admission, Mr Xiberras had been exerting such pressure upon the Gibraltar Government both before and after the last general election in Gibraltar of June 1972. Secondly, it could be contended that the two page feature article on Gibraltar published in *The Times*[12] acted as a spur for Sir Joshua's original letter and that the reply to that letter by the Spanish Ambassador in London forced Sir Joshua's hand and obliged him to make public the contents of the hitherto undisclosed Spanish proposals. Such an argument, however, implies that Sir Joshua was sufficiently naive to be drawn in that way; something which is hard to credit given his prominent, lengthy and remarkably successful involvement in Gibraltarian politics. Thirdly, it may be argued that Sir Joshua quite deliberately revealed the Spanish proposals when he did as a means of exerting pressure on the British Government to increase its financial support for a Gibraltarian economy which was deeply depressed by the continuing Spanish frontier restrictions.

This last suggestion may well afford the best explanation of this otherwise curious and largely inexplicable twist in the Gibraltar saga; especially when it is noted that Sir Joshua and the Governor of Gibraltar, Sir John Grandy, were in London at the head of a Gibraltar delegation at the time that Sir Joshua's revelations appeared in *The Times*.[13] In July, at the general assembly of the Integration With Britain Party, Mr Xiberras had exhibited deep concern at the rapidly rising cost of living in Gibraltar and had called for a constitutional conference to be held with the objective of establishing a 'special economic relationship' between Gibraltar and Britain. Similarly, shortly before leaving for London, Sir Joshua had also revealed his thinking on the needs of Gibraltar's economy. He expressed the hope that during his talks with representatives of the British Government he would be able to negotiate a three-year aid programme worth £12 milllion to Gibraltar. Such a programme would reflect both Gibraltar's high rate of inflation and the constraints imposed upon the city's economy by the continuing failure of the Anglo-Spanish negotiations to resolve the Rock's future status and to lift the frontier restrictions imposed by the Spanish Government. It was Sir Joshua's belief that these factors, collectively, necessitated a change in the emphasis of Britain's aid programme from 'temporary measures' designed to sustain the Gibraltarian economy in the short term to investment in larger projects which would increase the Rock's self-sufficiency in the long term.

If the parlous state of Gibraltar's economy accounted for Sir Joshua's disclosure of the Spanish proposals and if that disclosure was timed to secure further British aid for the Rock's economy then it was to prove to be a remarkably successful political ploy. On 13 November the Ministry for Overseas Development announced that Britain would provide capital aid amounting to £7,662,000 for Gibraltar's development programme for the three years 1975–6 to 1977–8 and would also continue to provide technical assistance with an increased provision for teacher training. The announcement of the aid programme was made by Mrs Judith Hart, Minister for Overseas Development, who said that she:

> accepted the need for aid beyond 1974–75 towards the development programme presented by the Gibraltar Government ... as a continuing fulfilment of [Britain's] undertaking to sustain and support the people of Gibraltar in the difficult circumstances caused by the Spanish restrictions.[14]

Unquestionably, the Spanish frontier restrictions had severely impeded Gibraltar's commercial and industrial life and occasioned the loss of the city's former Spanish workforce. Yet other factors also contributed to the damage sustained by the Rock's economy. Foremost among these was the rising cost of living; which had been highlighted earlier in the year by Mr Xiberras. As Gibraltar placed great reliance on imported commodities this meant that it was particularly vulnerable to worldwide inflationary trends and the index of food prices on the Rock was rising at an annual rate in excess of 20 per cent in the first half of 1974 while ttthe increase in the cost of oil had added an additional £500,000 to the year's budget of £7 million. Furthermore, in the early autumn, the Spanish newspaper, *La Vanguardia*, noted the worsening labour shortage in Gibraltar as many Portuguese and Moroccan workers, who had helped to fill the vacancies left by the former Spanish workforce, returned home having become discontented with high living costs, low wages and poor living conditions on the Rock.[15]

The rising cost of living was undoubtedly a contributory factor underlying the several weeks of industrial unrest which beset Gibraltar, immediately before negotiations over the new aid programme began, as trade unionists launched a campaign for an improvement in the 7 per cent increase in basic rates of pay offered by British and Gibraltarian government departments after the most recent biennial review of public-sector wages and salaries.[16] Nor was this the first wave of industrial unrest in Gibraltar associated with the biennial review, for the previous review had led to a general strike by industrial workers in the public sector which had severely disrupted public services and only ended with the Gibraltar Trades Council's (GTC) acceptance of the employers' offer of a minimum wage for both manual and white-collar workers.[17]

The 1974 biennial review of public-sector wages and salaries was complicated by a split which had developed within the trade union movement in Gibraltar. Previously, the Gibraltar branch of the British Transport and General Workers' Union (TGWU) had enjoyed a near-monopoly position in the representation of labour but, early in 1974, Mr Bernard Linares had founded the Gibraltar Workers' Union as a territorially confined alternative to the Gibraltar branch of the TGWU. Despite this schism in the ranks of organised labour, the leader of the TGWU, Mr Jose Netto, continued to campaign for parity of wages and salaries with workers in Britain[18] and, with a profound labour shortage on the Rock, the TGWU was in a strong position from which to advance its claim.

The Gibraltar Government, however, rejected the claim for parity because it held that such a concession would have necessitated average weekly wage increases of £12 for manual workers and £16 for white-collar workers; which would have cost in excess of £2 million per annum to implement. The Government argued that such an additional burden would be 'ruinous' if placed upon an already ailing economy. The GTC disputed the Government's figures and gave the annual cost of parity as £800,000 which it claimed was a reasonable price to pay for the elimination of the existing pay disparities between the local workforce and expatriates from Britain; a disparity which had long been a source of grievance. The existence of those disparities was denounced by the GTC as a reflection of 'the continuation of colonialism in Gibraltar'.

Negotiations between management and labour broke down and, at the beginning of October, employees in government departments introduced 'go-slow' practices and were supported by the GTC which issued a warning that it was prepared to cut off power supplies at fifteen minutes' notice if talks were not resumed. Later in the month a delegation from the GTC visited London and agreement was reached on the negotiation of an interim pay award to be followed by an inquiry into the problems underlying the dispute over pay parity with Britain.

The employers' interim pay offer, which amounted to weekly increases ranging from £3.40 to £6, was rejected by the GTC which demanded an across-the-board increase of £10; in support of this demand the 'go-slow' was re-imposed. The employers responded by laying-off some of those involved in this latest round of industrial action and the GTC responded by calling a one-day strike and demonstration on 4 December.

After the events of 4 December an uneasy truce developed while an investigation of the parity issue was undertaken by a team headed by Sir Jack Scamp. The team concluded that:

> as a guiding principle the parties should aim to establish a more stable relationship between Gibraltar and UK wage and salary rates approximate to 80 per cent of the UK rates for corresponding grade of employees. This relationship should be phased into full effect progressively.

Another sixteen months was to pass until, in May 1976, the Gibraltar branch of the TGWU agreed to accept an offer based on the Scamp proposals and before further agreement was forthcoming that 100 per cent parity could be sought in future negotiations.[19]

It was perhaps inevitable that, following the furore occasioned by Sir Joshua Hassan's revelations of 1974, with the frontier restrictions imposed by the Spanish authorities still in place and with Anglo-Spanish negotiations on the Gibraltar question in abeyance, 1975 would prove to be a year in which the Rock and its future assumed a lower priority in both London and Madrid. This otherwise tranquil interlude was disturbed somewhat in September when Mr Roy Hattersley, Minister of State at the Foreign and Commonwealth Office, paid an official visit to Gibraltar to familiarise himself with the situation.[20] The timing of the Minister's visit was perhaps unfortunate in that it coincided with the work of a Constitutional Committee, comprising representatives of both the Gibraltar Government and Opposition, and also took place during the preliminary stages of a campaign leading to a general election in the city. At the end of his visit, Mr Hattersley ruled out any future prospect of the complete integration of Gibraltar and Britain whilst stressing that the British Government would continue to respect those provisions contained within the preamble to the Gibraltar Constitution which precluded the transference of sovereignty to another nation against the expressed wishes of the Rock's population. Mr Hattersley made his position clear on these points at both a press conference and in an interview screened on Gibraltar television.

Mr Hattersley's publicly expressed remarks led to a letter from the Leader of the Integration With Britain Party, Mr Maurice Xiberras, to the Governor of Gibraltar, Sir John Grandy, in which he protested at Mr Hattersley's interference in the domestic affairs of the Rock. In reply, Sir John contended that Mr Hattersley's comments were personal rather than official in nature and had no bearing on the forthcoming election.[21] The Integration With Britain Party, which had been specifically formed as a reaction against the intensified Spanish campaign to recover sovereignty over Gibraltar, responded to Mr Hattersley's rejection of the Rock's integration with Britain by walking out of the House of Assembly and remaining away throughout the second half of October.

Mr Xiberras was to justify his Party's departure from the House of Assembly by contending firstly, that the Integration With Britain Party had never proposed to the Constitutional Committee 'complete integration' with Britain, if that was defined as representation in Westminster, because it recognised the difficulties that would be involved in gaining acceptance for that concept not only in Britain but also in Gibraltar. Secondly, that the Integration With Britain Party's:

> protest in the House of Assembly was directed at the general impropriety of [HMG] making the kind of general pronouncement, which the undiscriminating in Gibraltar might take to be a general condemnation of our party, and quite gratuitously at that, because Mr Hattersley was aware from our talks with him that our constitutional proposals did not involve complete integration. Our protest was directed at Mr Hattersley's interference with local party politics in a pre-election period.[22]

However, of more immediate significance to Gibraltar's future was the death of General Franco on 20 November and the accession, two days later, of King Juan Carlos to the Spanish throne.

This change in the political regime of Spain occasioned hopes that a different attitude to the Gibraltar question would be adopted in Madrid despite the fact that the King had emphasised, in a speech marking his ascendancy to the throne, that Spain would continue its efforts to regain sovereignty over Gibraltar. Such hopes were encouraged when the newly enthroned monarch ordered that telephone links between Gibraltar and the Spanish mainland, which the Spanish Government had closed in October 1969, would be re-opened over the coming Christmas period.[23]

Caution, however, continued to characterise the Spanish Government's approach to the Gibraltar issue. The new regime's Foreign Minister, Señor Jose Maria Areilza, declined, in early March 1976, to commit himself on whether Spain was willing to make a token relaxation of the restrictions on Gibraltar in order to improve the climate for a resumption of Anglo-Spanish negotiations. He stated that he was prepared to discuss Gibraltar 'but the way I do it depends on the echo I find in the British Government'.[24] Señor Areilza had an early opportunity to test the resonance of the echo reverberating from the British Government when he made an official visit to London for talks with Prime Minister Wilson and Foreign Secretary Callaghan later that same month. Confronted by Mr Callaghan's re-statement of the long-standing British position that there could be no change in the sovereignty of Gibraltar without the consent of the Gibraltarians themselves, Señor Areilza conceded that the newly democratised Spain might assume a more flexible attitude to the Gibraltar question than had been possible under General Franco but he stopped short of granting any relaxation of the frontier restrictions imposed during the Caudillo's rule.

Yet hopes for a workable settlement to the Gibraltar question remained buoyant, especially among those who saw a solution lying in both Britain and Spain being fellow members of the EC, since Señor Areilza had publicly expressed the view that whilst Spain had not asked for formal negotiations on its entry to the Community to begin, he held out the prospect that a positive reaction to the Spanish membership, both within and without Spain, was emerging. During his talks in London, Señor Areilza had been assured by his British counterpart, Mr Callaghan, that Britain would welcome a Spanish application to join the EC because a democratic Spain would be a great addition to the strength of southern Europe.

The freshly emerging warmth in Anglo-Spanish relations had an impact upon the political life of Gibraltar. Just as had happened in 1972, the general election of 1976 was dominated by a single issue: that of Gibraltar's future status. Unlike 1972, however, the range of scenarios for the Rock's future status was now far greater.

This provided an unheralded number of candidates for an election in Gibraltar, where the more customary pattern had been eight candidates against eight, drawn on clear political party lines, for the fifteen elected seats. Traditionally, this had resulted in only one candidate failing to gain election and the formation of an eight-man Government, all holding ministerial office, and a seven-man opposition, all serving as shadow ministers.

In the initial run-up to the 1976 election it had appeared that the Gibraltarian electorate would be presented with a choice from three differing options relating to the Rock's future: integration with Britain; the customary demand, association with the political left, for liberation from colonial rule; and a cautious approach to Spain. However, the first of these three options was removed from the hustings when Mr Maurice Xiberras resigned as leader of the Integration With Britain Party on 7 September. With his resignation the party was effectively dissolved. The Integration With Britain Party had formed the Opposition during the last House of Assembly but had steadily lost ground since Mr Hattersley's fact-finding visit to Gibraltar a year earlier when he had ruled out the possibility of the complete integration of Gibraltar with Britain. More recently, during talks in London plans for constitutional reforms in Gibraltar involving greater integration with Britain had been rejected by the British Government as 'neither desirable nor practicable'.[25] Collectively, these two rebuttals had sounded the death knell of the Integration With Britain Party and Mr Xiberras's resignation sealed its fate.[26] Mr Xiberras's political career, however, did not end for he was to successfully contest the election as an Independent.

A new political grouping, the Gibraltar Democratic Movement, headed by Mr Joe Bossano, a prominent and long-serving trade union leader, stepped into the void left by the demise of the Integration With Britain Party. Mr Bossano had been a founder member of the Integration With Britain Party, which he had left in November 1975, but he had remained in the House of Assembly as an Independent. The formation of the Gibraltar Democratic Movement could be directly attributed to the refusal of the British Government to discuss a new constitution for Gibraltar with anyone except the elected representatives and the Gibraltar Democratic Movement stated in its election manifesto that 'the present situation in Gibraltar is an affront to the Gibraltarian people' and pledged the fledgling Party to 'work for the decolonisation of the Rock and the creation of a new constitutional arrangement which will guarantee the future of the territory and the people'.

Of the eleven independent candidates seeking election, three favoured closer links and negotiations with Spain. Essentially, they advocated Gibraltarian representation at early negotiations between Britain and Spain to guarantee the Gibraltarian way of life; a re-opening of the frontier with Spain; and a return to normal social and economic relations across the frontier. They viewed these prospects as distinct possibilities as Spain moved towards increased democracy and membership of the EC.[27]

The out-going Chief Minister, Sir Joshua Hassan, again led his party into the elections and its platform, with emphasis heavily slanted towards Gibraltar's domestic affairs, stood in marked contrast to that of its rivals.[28] None the less, Sir Joshua was adamant that he would not be hurried into any early negotiations with Spain as it began to develop democratic institutions and political processes, nor was he inclined to exert pressure on Britain to grant Gibraltar what he contended would be a short-lived independence. Instead, he reasoned that, in a changing Europe, Gibraltar would eventually have to live on normal terms with Spain and that he would then seek to gain Spanish recognition for the existence of the people of Gibraltar as a separate community. In short, he adopted the middle ground on proposals concerning the Rock's future status and added to this pragmatic posture a domestically oriented balance.

As the final few days of campaigning took place in Gibraltar the new Spanish Foreign Minister, Señor Marcelino Oreja Aguirre, repeated Madrid's demand for the return of the Rock to Spanish sovereignty. When the general election took place on 29 September, the Association for the Advancement of Civil Rights secured the necessary one-seat majority to form the Rock's next Government and Sir Joshua was again re-appointed to serve as Chief Minister. The Gibraltar Democratic Movement, contesting its first election, secured four seats with the three remaining places on the Opposition benches going to Independents, all of whom had served in the last House of Assembly as representatives of the Integration With Britain Party. In a turn-out of 75 per cent some 14 per cent of the voters had given their support to the three Independents who had campaigned on a platform of accommodation with Spain – insufficient to assure the election of any one of their number.

Despite the domestic changes taking place in Spanish politics, despite the growing accord in Anglo-Spanish relations, the electoral outcome could only be regarded as an expression by the people of Gibraltar that they were neither willing nor prepared to be hastened into any settlement with Spain and that they continued to see their future as inextricably interwoven with Britain.

7

HOPE IS SUCCOURED

The discernible movement towards the development of democratic institutions and democratic political processes in Spain and the parallel growth in the cordiality of Anglo-Spanish relations which had been evidenced since the death of General Franco in November 1975 had failed to produce any alleviation of the Spanish pressure upon Gibraltar. The frontier restrictions which had been progressively imposed by the Spanish authorities were still in place and it seemed that there was no immediate sign of positive progress towards a settlement of the Gibraltar problem when, at a ceremony in honour of the late Spanish Foreign Minister, Señor Fernando Maria Castiella, held in San Roque, the new Spanish Foreign Minister, Señor Marcelino Oreja, promised to seek a formula that would return Gibraltar to Spain.[1]

A month after Señor Oreja's speech, in March 1977, the findings of a private study of Gibraltar were published which concluded that integration with Spain offered the Rock the best long-term opportunity for economic development. The study recognised that any future change in Gibraltar's status was dependent upon fundamental political changes taking place in Spain and conceded that Spain had not yet progressed far enough along the path to democracy for complete integration to be a current option. However, it was emphasised that should integration not take place then the scenario for Gibraltar's future was bleak – conditioned as it was by geographical isolation, dependence on unpredictable external factors, a stagnant economy, the exodus of young people and with the prospect of more serious social and economic problems pending. The study surmised that these factors, coupled with the dependency of the Gibraltarian economy upon a British military presence, worth £8 million, and uncertainty about Britain's future defence expenditure, indicated that the Rock's future lay under Spanish sovereignty.[2] This report fuelled speculation in Gibraltar, which had been rife at the time of the last general election, that London and Madrid were moving towards a bilateral agreement on the Rock's future with Gibraltarian interests neither being represented nor given sufficient credence.

After a thirteen-hour Cabinet meeting on 12 July, the Spanish Government announced a wide range of far-reaching political and economic policy decisions. The most significant of these for Gibraltar was the re-assertion that one of Spain's foremost political aims would include respect for the integrity of the national territory.[3] Undoubtedly, this referred, above all else, to the Rock, especially as it was also stated that the Spanish Government was again ready to re-open talks with Britain over the hitherto unresolved problem that Gibraltar represented. Nor were Gibraltarian anxieties allayed when the Spanish Government formally submitted an application for full membership of the EC on 28 July; thus becoming the third southern European country to do so.[4] The Spanish application was welcomed by the Community's Foreign Ministers on 20 September and, following this, the European Commission was instructed to prepare a detailed report on the implications of full membership before final negotiations could begin.

To those in Gibraltar, like Chief Minister Sir Joshua Hassan, who had anticipated the eradication of frontier restrictions and a settlement of the Rock's future status being facilitated by Spanish entry into the EC, there were worrying signs much in evidence. Concern over the long-term viability of the Rock's economy were compounded by the re-iteration of Spain's intentions to restore its sovereignty over Gibraltar, and it appeared feasible that the Spanish Government's application for full membership of the EC, while causing the abandonment of the siege to which the city had been subjected, might also lead to the end of British sovereignty being negotiated between London and Madrid. Equally disturbing to many in Gibraltar was the fact that the British Government was disinclined to exert pressure on the Spanish Government by making its support for Madrid's application to join the EC conditional upon Spain ending the frontier restrictions which it had imposed upon the Rock and which had progressively led to the parlous state of the city's economy. That this was the case was evidenced at the end of a visit made by the British Foreign Secretary, Dr David Owen, to Madrid, for talks with his Spanish counterpart, Señor Oreja, and Prime Minister Adolfo Suarez.[5] During those talks Dr Owen had asked the Spanish Government to lift the frontier restrictions imposed since 1969, namely, the closure of the Spanish–Gibraltar border, the suspension of the Algeciras–Gibraltar ferry and the severance of telephone links between Gibraltar and the Spanish mainland. However, Dr Owen, was at pains to point out that British support for Spain's application for membership of the EC was not conditional upon a settlement of the Gibraltar question and that British support for Spanish membership of NATO would also be forthcoming should Madrid decide, after parliamentary debate, to apply. He added that he had not gone to Madrid expecting immediate results on the Gibraltar question but that he had found 'a degree of sensitivity and understanding which did not exist before and which was the best ingredient for a settlement'. As had become *de rigueur* for all British Foreign Secretaries when pronouncing on Gibraltar, Dr Owen stated that the only feasible settlement of the question posed by the Rock would be the one which took into account the wishes of its people.

The uncertainty surrounding the eventual outcome of those wider developments had an impact upon the domestic politics of Gibraltar. Despite the growing vociferousness of both the Government and Opposition in their re-iteration of their strong pro-British sentiments, other factions, such as the newly-formed Party for the Autonomy of Gibraltar, were beginning to press the case for a settlement with Spain beyond the confines of the House of Assembly. In particular, the Party for the Autonomy of Gibraltar, called for an agreement between Britain, Spain and Gibraltar to afford the Rock autonomous status under the new Spanish constitution; a status akin to that being considered in Madrid for the Basque and Catalan provinces.[6]

There was speculation in the Spanish media during the early autumn that the frontier restrictions on Gibraltar might be removed by Christmas but it was stressed in Spanish governmental circles that a unilateral act was not to be anticipated and that 'the relaxation of the so-called border restrictions should be considered within the framework of Spanish–British relations about the decolonisation of Gibraltar'. Media hypothesising aside, it was evident that the new regime in Spain was producing governments that were prepared at least to talk about negotiations with Britain, and evidence of this trend was forthcoming when Señor Oreja told the General Assembly of the UN that Spain would lift the restrictions imposed upon

Gibraltar after the resumption of talks with Britain and that 'we will take into account the interests of the inhabitants [of Gibraltar] and facilitate movement between the mainland and the surrounding areas in order to allow progress in the negotiations'. Señor Oreja went so far as to suggest that formal talks on Gibraltar would begin soon after a forthcoming visit to London by the Spanish Prime Minister.[7]

Shortly before Señor Oreja had departed for the UN in New York there had been a debate on foreign policy in the Cortes.[8] From that debate there emerged unanimous agreement on the desirability of Spanish membership of the EC and upon the 'decolonialisation' of Gibraltar. Almost equal unanimity was reached when the question of Spanish membership of NATO was discussed except, on this question, the idea of membership was rejected by all political parties save for the Popular Alliance (UCD); the UCD, however, advocated further debate on this topic. Two days before the Spanish Prime Minister left for London, for talks arranged during Dr Owen's visit to Madrid, a group of about 300 Spaniards shouting 'Open Up' demonstrated at the closed frontier gates at the entrance to Gibraltar from the Spanish mainland. They were joined by roughly 100 Gibraltarians who echoed the Spaniards demands. The Spanish demonstrators carried a placard signed by a 'Committee for Reconciliation' which had been formed a fortnight earlier by residents from the localities adjoining Gibraltar. As has been mentioned previously, Spanish frontier restrictions had not only had an adverse effect upon the Gibraltar economy but also upon that of the towns of the Campo region.[9]

Two seemingly contradictory signals, namely, the total accord reached in the Cortes on the need for the 'decolonialisation' of Gibraltar and the demands voiced by some of those Spaniards living in the Campo region for the ending of frontier restrictions on Gibraltar, were communicated to Señor Suarez as he prepared to depart for London for talks with his British opposite number, Mr James Callaghan, and Foreign Secretary, Dr David Owen.[10] At the conclusion of his visit Señor Suarez stated his belief that Spain's political evolution could permit a negotiated settlement to the Gibraltar dispute. Significantly, he made reference to a solution in the context of a 'regionalised' Spain:

> there is no doubt that Spain, which, is being structured on a regional basis with respect for and recognition of, autonomy for each of the peoples which make up the Spanish state, can

hold out the hope of a negotiated statute respecting the identity, culture and special characteristics of the Gibraltarian people, and eventually can bring about the reintegration of Gibraltar into Spanish territory in conformity with UN resolutions.

However, Señor Suarez rejected the possibility of removing the frontier restrictions imposed by the Franco regime and said that any agreement on the Rock's future would have to be 'global' rather than arrived at step by step. As Señor Suarez departed for Madrid a communiqué was issued from Downing Street which emphasised that for there to be progress on the question of Gibraltar the frontier restrictions must be lifted. Again the Gibraltar problem had proved to be intractable.

In keeping with the pattern established after previous Anglo-Spanish talks in London, Señor Suarez's departure was shortly followed by a visit from representatives of the Gibraltar Government to the Foreign and Commonwealth Office.[11] On his return to Gibraltar, Sir Joshua Hassan told the House of Assembly that he had proposed to the British Foreign Secretary that direct talks should be held between the British and Spanish Governments, with Gibraltarian representation, as part of a major initiative to break the continuing impasse. Sir Joshua was adamant that the talks should be exploratory in nature and without commitment on either side. The Chief Minister enjoyed the unanimous support of the House of Assembly for his proposal.[12]

Sir Joshua's desire for exploratory and non-committal talks involving representatives from the three governments involved in the problem of Gibraltar was fulfilled on 24 November in Strasburg. Sir Joshua and the Leader of the Opposition in the Gibraltar House of Assembly, Mr Maurice Xiberras, represented the Rock as part of a British delegation led by Dr David Owen. After a meeting of some two and half hours with a Spanish team headed by Señor Oreja a communiqué was issued which described the meeting as having been 'friendly, co-operative and constructive'; made mention of a further round of Anglo-Spanish talks to be held in 1978; and stressed the need for 'good faith and realism'. Yet below the veneer of cordiality it would appear that no movement had occurred nor progress made and that well-prepared and firmly entrenched positions had been used from which to fire off the old, familiar demands. There was no indication that the Gibraltar representatives had suggested that the people of the Rock were any more ready to abandon their opposition to Spanish rule despite the growth of factions favouring association with Spain, nor did Dr Owen do otherwise than emphasise the 'exploratory and non-commital' character of the talks; whilst Señor Oreja gave no indication that Spanish frontier restrictions would be lifted in the foreseeable future. However, Señor Oreja gained what had been the primary objective of his visit to Strasburg; not a settlement of the Gibraltar dispute but Spanish membership of the Council of Europe.[13]

As Christmas 1977 approached, the Spanish authorities again re-connected the telephone lines between the Spanish mainland and Gibraltar but, unlike a year earlier, telephone communication was not severed in the new year and thereafter the phone lines remained open and Gibraltar's isolation was marginally lessened.

The new year saw apparently contradictory signals being flashed from Madrid: signals which gave the appearance that the Spanish Government was unsure what moves were best suited to its objectives over Gibraltar in the more constructive atmosphere of the 'new spirit' which had been established at Strasburg.[14] In early January, Señor Oreja told the Spanish Parliament that there could be no question of Spain lifting the frontier restrictions which had been imposed upon the Rock and that Britain continued to defy UN resolutions which had called for the decolonialisation of Gibraltar.[15]

Yet, within a few days, Señor Javier Ruperez, who was in charge of the foreign policy of the Centre Democratic Union, the party of Prime Minister Suarez, was travelling to Gibraltar, via Morocco, for talks with the Rock's political leaders.[16] Señor Oreja's hardline on Gibraltar of early January appeared to have undergone a profound transformation by the end of that month. Whether this was as a result of Señor Ruperez's discussions in Gibraltar is impossible to discern but it would seem that those talks must have had some bearing on the change which had occurred for, on 30 January, the Spanish Foreign Minister was to tell foreign correspondents at a lunch in Madrid that the Spanish Government was 'disposed to recognise Gibraltar's special identity once the colonial process had been terminated'; that Britain could continue to use the naval base at Gibraltar if it accepted Spain's claim to sovereignty over the Rock; and that he would contemplate 'an even broader formula' for regional autonomy for Gibraltar than had been granted to other Spanish regions. He did, however, point out that the subject of a regional statute for Gibraltar had not been discussed in meetings with representatives of the British and Gibraltarian Governments.

Perhaps most significantly, Señor Oreja made a radical departure from the argument that had been pursued with such vigour by former Spanish foreign ministers when he said that: 'I do not think that either Spain or Britain should tie itself down to a treaty which was signed in 1713. Spain prefers to look upon Gibraltar in a present day context and not in terms of the Treaty of Utrecht, if that is possible'.[17]

Yet it had been the Treaty of Utrecht which had previously been cited by Spanish Government representatives as the basis for Spain's claim to sovereignty over the Rock and which had been employed so effectively at the UN to mobilise international support against the position adopted by successive British Governments on the question of Gibraltar. This shift, from recourse to the past to emphasis upon 'the present-day context' could be taken to imply that the Spanish Government now saw a resolution of the Gibraltar question lying in the wider role that Spain wished to play in the EC. Further, it could be read as indicating that once Spain had joined Britain as a member of the Community the growing commonality of the interests of the two nations that would inevitably ensue would outweigh that which manifestly divided them: namely, Gibraltar, and that reason would dictate that a settlement of an old dispute would have to be forthcoming if anticipated, mutual benefits were to be secured.

The new, softer Spanish position on Gibraltar was to be quickly followed by an event which was to embarrass the Spanish Government in its claim to sovereignty over the Rock. A meeting of the Council of Foreign Ministers of the Organisation of African Unity (OAU), in Tripoli, affirmed the 'African character' of the Canary archipelago and called upon African states to lend moral and material support to the Canary Islands Independence Movement (MPBIAC).[18] It would appear probable that the OAU voted in support of the MPBIAC because of Spanish support for Morocco which was, together with Mauritania, engaged in resisting the guerilla tactics employed by the Polisario Front in pursuit of its claim for independence from both Morocco and Mauritania in what had formerly been Spanish Sahara.[19] The response from Madrid to the OAU resolution was predictable with the Spanish Government vigorously condemning 'a proposal which constitutes an intolerable interference in Spanish internal affairs'.[20] Although the questions of independence for the Canaries and of the future status of Gibraltar were totally discrete issues they could be seen to be closely connected, for if the people of the Canaries had the right to remain a part of Spain, then so too did the inhabitants of Gibraltar have the right not to become a part of Spain; geographical proximity could not be permitted to override the wishes of the people most directly affected. To compound the Spanish Government's embarrassment, King Hassan II of Morocco was openly to draw a parallel between Gibraltar and the two Spanish enclaves in Morocco, Ceuta and Melilla, when he noted that these would have to be returned to Morocco if Gibraltar was to be returned to Spain.[21]

The next round of Anglo-Spanish talks on Gibraltar's future, which it had been agreed to hold when the parties had last met in Strasburg, took place in Paris on 15 March. Again, Gibraltar was represented by its Chief Minister, Sir Joshua Hassan, and the Leader of the Opposition, Mr Maurice Xiberras. The 'new spirit' established at Strasburg was once more much in evidence although it was recognised that attitudes which had been developed over decades could not readily be changed and that the major problem of reconciling the strongly held Spanish claim to sovereignty over Gibraltar, and the Gibraltarians' equally strongly held desire for a continuation of their independence from Spain still stood in the way of a speedy settlement. Nevertheless, the 'new spirit' undoubtedly facilitated a modicum of progress as was attested by the creation of three joint working parties to examine the re-opening of the Algeciras-Gibraltar ferry service, the permanent restoration of telecommunications between Gibraltar and the Spanish mainland, and the payment of social security benefits to those Spaniards who had lost their jobs in Gibraltar when the frontier was closed. The resolution of all of these issues was seen as being a necessary prerequisite to the rebuilding of mutual confidence between Gibraltar and Madrid. Both British and Spanish representatives stressed that the problem posed by Gibraltar to their future accord was now being viewed very much in the 'present-day context' of which Señor Oreja had previously spoken, and that important elements in that context were the Spanish application to become a partner to Britain in the EC and deliberations then taking place in Madrid with a view to Spain seeking membership of NATO.

The Spanish Parliament had begun to debate its country's possible entry into NATO in March, at a time when the Spanish left had again started to reiterate its opposition to such a commitment. Both the Socialist and Communist parties had anti-Americanism as one of their most deeply rooted tenets which had arisen as a reaction to American influence, both upon Spanish domestic and foreign policy, during the Franco regime. Whilst there had been unanimous political party agreement to Spain's application to join the EC and whilst two of the most senior figures in the Spanish Government, the Prime Minister and the Foreign Minister, favoured Spanish membership of NATO the opposition of the left seemed guaranteed to ensure that there could be no consensus on the NATO question. The view of the Spanish Government was that Spain already had contractual obligations to the West and, were Spain to adopt a neutralist stance to East–West antagonisms, this would have adverse repurcussions on the balance of power in Europe.[22]

However, whilst Western estimates showed the numerical strength of NATO to be well below that of the Warsaw Pact it is difficult to see how the mere addition of Spanish military forces would have made for any real redress of the imbalance; but Spain's strategic position, as a guardian of NATO's southern flank and overlord of the seaways between the Mediterranean and the Atlantic Ocean, could be viewed as making a significant contribution to the overall European balance of power between East and West.[23] Despite the arguments that could be mustered by all of the parties in the Spanish Parliament as to the rights and wrongs of Spanish membership of NATO one factor upon which they were all agreed, and upon which they were all united in perceiving as a real barrier to Spanish membership, was the Spanish claim to sovereignty over Gibraltar. Spain could hardly become a member of an international military alliance when one of the founder members of that alliance, Britain, 'owned' a piece of territory that was agreed by all shades of Spanish domestic political opinion to be an integral part of Spain. Thus the Gibraltar question not only exerted a direct influence upon Anglo-Spanish relations but also served as a constraint upon the wider field of Spanish foreign policy.

Britain, as the other major party to the question of Gibraltar's future status, was also discovering that the Rock had a propensity to exert an influence beyond the narrow confines of Anglo-Spanish relations. Three areas of British foreign policy seemed to be affected by the inability to reach a negotiated settlement of Gibraltar's future. It would be unrealistic to contend that each of these carried equal significance for the the foreign policy makers in Westminster and Whitehall.

The first, and certainly the least important, was the question of an official visit by King Juan Carlos to Britain. The King had made a private four-day visit to Britain in January 1978 but Britain had remained noticeably absent from the growing list of official visits he had made abroad and which had been made to Spain by other heads of state since Spain became a democracy. However, were such a visit to be made by the Spanish monarch to Britain this would unquestionably have been met with a hostile reaction in Gibraltar where cordiality to the head of a nation that was causing such prolonged hardship to that small community would have been likely to have led to a deterioration in relations between Gibraltar and London, thus making the problem of Gibraltar even more intractable – if such a thing was possible – than it had been so far. Additionally, the Gibraltarians would have been quick to remind the British Government that two visits to the city by members of the British Royal Family had been the subject of official protests from the Spanish Government.

Secondly, and of greater political moment, was the question of British support for Spanish entry into the EC. To date Britain had shown a readiness to support the Spanish application and had resisted threatening the withholding of support as a way of bringing pressure upon Spain to remove the frontier restrictions it had imposed upon Gibraltar or as a means of persuading Madrid to accept the continuation of British sovereignty over the Rock. It seemed unlikely that Britain would implement such a threat, especially as Sir Joshua Hassan had said as recently as May 1978 that he saw the best hope for a solution to Gibraltar's future lying in the context of Europe.[24]

Thirdly, and of most political importance, was the prospect of Spain applying to become a member of NATO. With Britain effectively having long withdrawn from its 'policeman' role east of Suez the military importance of Gibraltar to Britain as a base, from which to guarantee its seaways to those areas formerly of importance to the preservation of her national interests in the Middle and Far East, had declined. The military diminution of Gibraltar in British defence policy can be gauged by the fact that Britain had not installed the most technologically modern military hardware and software on the Rock; this perhaps reflected both Gibraltar's declining status as a British military base per se and British concern that Gibraltar was a base which could be disrupted by Spain. Yet, a continued British military presence in Gibraltar was vital to preclude the Rock's strategic position being used by other unwelcome powers; to sustain its economy, which was heavily dependent upon continued British military expenditure and aid; and as a means of being seen to be honouring the undertakings given to the city's inhabitants by successive British Governments over many years.[25]

Additionally, it cannot be discounted that Britain may well have been subjected to increasing American pressure to resolve the Gibraltar dispute; not simply because of traditional US anti-colonial rhetoric but also as a means of facilitating Spanish membership of NATO. As long ago as May 1975, US President Gerald Ford and Secretary of State, Dr Henry Kissinger, had met with General Franco and, the then Prince, Juan Carlos, in Madrid after the spring meeting of the North Atlantic Council in Brussels. At that time President Ford had observed that:

> Spain is part of the Atlantic Community. I have no doubt of the increasingly important role that Spain will play in the future of the world as a whole, and particularly in the West. I am sure that the future holds for Spain a greater role in the European and Atlantic organisations.[26]

The future for Spain envisaged by President Ford now appeared to have become the present, and brought with it its associated difficulties for British foreign policy.

For the immediate present there was cause for optimism as there appeared to be agreement in both Britain and Spain that the frontier restrictions would have to be removed if fruitful negotiations over Gibraltar's future were to take place. After all, the three working parties established at the Paris talks had been charged with investigating differing aspects of that very issue but at their first meeting, in London, they had failed to make any positive headway.[27] That progress was still not forthcoming on ending the 'siege' of Gibraltar may be attributed to the limited room for manoeuvre afforded to both the British and Spanish Governments. Spain had made clear its abiding desire to restore its sovereignty over Gibraltar; Britain had responded by refusing to adopt any course of action contrary to the wishes of the Rock's inhabitants and, in consequence, was at least publicly precluded from discussing the question of sovereignty. For their part, the Gibraltarians had repeatedly stated their preference to maintain their links with Britain, whilst not dismissing a new relationship with London which would afford them greater independence, and their opposition to Spain regaining sovereignty over their city.

In this context the questions confronting all three actors in the on-going drama were; what substantive concessions, short of the restoration of Spanish sovereignty over Gibraltar, could Britain make, and Spain accept, in return for lifting the frontier restrictions? What could the Gibraltarians do to ensure that Britain and Spain respected their wishes as to their future status whilst at the same time preventing London and Madrid entering into negotiations leading to outcomes unacceptable to the Rock's community? Within these parameters the debate on Gibraltar's future was set to continue.

If matters European were to provide the context for the resolution of the Gibraltar problem then 1978 offered initial signs of promise. On 20 April the Council of Ministers received from the Commission three reports arising from the membership applications tabled by Greece, Portugal and Spain. These reports focused on: the implications of enlarging the Community; the transitional period from application to full membership and institutional arrangements; and economic and sectoral issues. The Commission viewed all three applications as having arisen from political considerations by each of the applicant countries in that all of them had been primarily motivated to seek membership of the EC as a means of fostering their recently introduced democratic institutions. It soon became clear, however, that the third report of the Commission, namely, that concerned with the economic and sectoral issues of an enlarged Community, was to be the subject of heated debate; a debate which was to be continued throughout the year. Specifically, influential political opinion in both France and Italy had expressed concern at first, and later opposition to, Portuguese and Spanish membership because of the likely envisaged damage that would be inflicted upon the French and Italian agricultural sectors.[28] On 19 December the Council of Ministers decided that a ceremony to mark the start of accession negotiations with Spain would be held in the spring of 1979.

However, at the insistence of France, the member nation which had been most vociferous in its opposition to the admittance of both Portugal and Spain to the Community, it was decided that substantive negotiations would not begin on the Spanish application until representatives of the EC and the Spanish Government had agreed a common negotiating basis; a process which would inevitably delay Spanish entry. It thus began to appear that if a settlement of the Gibraltar dispute was to be forthcoming upon Spain's entry into the EC then that settlement would not be effected in the immediate future. Yet at the same time it also began to appear that, as a result of French opposition, Spain might have greater need of British support to secure its admittance to the EC and might, in consequence, have to yield on Gibraltar by at least going so far as to remove the frontier restrictions.

After the general election of May 1979 a Conservative Government took power in Britain. This change in the political colour of the British Government did not at first appear to have any immediate implications for Gibraltar's future status, for the new Foreign Secretary, Lord Carrington, was quick to emphasise that the in-coming Government would continue the policies followed by the previous Labour Administration. However, within a month of assuming office, Lord Carrington told the House of Lords that the Government was to review its policy on Gibraltar and that early efforts would have to be made to settle the dispute with Spain.[29] The Governor of Gibraltar, General Sir William Jackson, sought to reassure the Gibraltarians that any review of British policy towards the Rock would continue to safeguard their future well-being. He expressed the view that the newly elected Conservative Government would develop a more rational policy towards the Rock and its inhabitants than that taken by its Labour predecessor, 'with the new Government it will be a much steadier business and I think it will be more effective'.[30] The Governor also seemed to rule out any transference of sovereignty from Britain to Spain when he stated that he felt it important that he should inform the Foreign and Commonwealth Office officials in Whitehall of the views of the Gibraltarians:

> I've got to keep saying to them, now come on, you look at Gibraltar as Gibraltar and its people, who are determinedly British, and you can't go with a great theoretical sweep of the hands and say Gibraltar is part of the Iberian peninsula and, therefore, should one day be part of Spain.

The British Government's review of its policy towards Gibraltar did not appear to produce any outward signs of change in emphasis. Lord Trefgarne was to tell the House of Lords, on 24 July, that, whilst it was inconceivable that the frontier restrictions could remain in force once Spain had joined Britain as a member of the EC, the British Government did not link Spanish accession with the lifting of those restrictions.[31] In short, it looked as though the policy review had failed to produce any new and alternative proposals on the Gibraltar question.

If the British Government could maintain continuity in its policy towards Gibraltar this was undoubtedly due to the absence of domestic pressure upon it to do otherwise. Such a luxury was not enjoyed by its Spanish counterpart for, in August, La Linea was again the scene of protests over Madrid's refusal to re-open the frontier with the Rock. An open letter to Prime Minister Adolfo Suarez, signed by 3,284 people living in the Spanish border town adjoining Gibraltar, asked for the blockade to be lifted. The letter contended that the policy of isolating the Rock, which had been pursued by successive Spanish Governments, had been a complete failure and called for a normalisation of relations between Spain and the City which, in the first instance, could be symbolised by re-opening all frontier crossing points once or twice per week. The letter also sought authorisation for local negotiations between Gibraltarian and Spanish representatives but stressed that the writers were not suggesting that Spain should abandon its claim to sovereignty over Gibraltar but were seeking an 'atmosphere of concord and friendship to help negotiations'.[32]

The first meeting between the recently appointed British Foreign Secretary and the Spanish Foreign Minister took place in New York on 24 September. At that meeting, British support for any forthcoming Spanish application to join NATO was promised: a timely decision, in that earlier in the month Señor Oreja had unequivocally stated that Spain would join NATO when the time was right and provided that the Spanish Government could secure the necessary parliamentary approval. However, the New York meeting failed to produce any progress on the Gibraltar question although both parties stressed that they were actively seeking a way of reconciling their differences.

It was the obvious lack of progress towards resolving the Gibraltar dispute which gave the impetus to new protests in the Campo region and which increased the domestic pressure that was being placed on the Spanish Government to lift the blockade. On 9 October, a 24-hour strike was staged in La Linea which brought industry, commerce and transport to a halt. La Linea, together with San Roque, the other town on the Spanish mainland closest to the Rock, had greatly suffered from the severance of its links with Gibraltar and approximately 4,000 of its insured population of 29,000 was unemployed as a direct result of the frontier restrictions.[33] In the Cortes, on 3 November, the Opposition Socialist Workers' Party

(PSOE) tabled a motion which proposed that the time had come for a debate on the ending of the frontier restrictions; it did, however, oppose linking the Spanish claim to sovereignty over Gibraltar with the negotiations which were then proceeding on Spain's application to join the EC or to any future discussion of Spanish membership of NATO.[34]

Clearly, with pressure building both within and without the Cortes, the Spanish Government had to respond and Señor Oreja did so in a policy speech made to the Foreign Affairs Committee of the Senate on 6 December. Whilst still demanding the restoration of Spanish sovereignty over Gibraltar the Foreign Minister took a softer line and, having described Gibraltar as 'an absurd anachronism', he conceded that 'the present situation cannot go on. The British authorities must accept the formal opening of negotiations'. This he felt the British Government would do, not just because of 'the pragmatic sense which the British have always demonstrated' but because new circumstances favourable to a satisfactory solution had come into being. He cited these as being: the new political structure in Spain with the possibility of regional home rule for Gibraltar; the emergence of an international climate in which nations with similar systems came together in pursuit of commonly held interests; and the realisation by the people of Gibraltar of their own uniqueness.[35]

Further evidence – of the growing urgency that was being increasingly felt in many quarters of Spanish political life, that the frontier restrictions which had been imposed upon Gibraltar were now proving to be counter-productive, had to be lifted and negotiations embarked upon – came the day following Señor Oreja's policy speech when a seminar organised by the Spanish Institute for International Affairs had approved a consensus which made certain proposals. These included: the presentation of a petition to the Spanish and British Governments urging them to take steps towards fostering a climate of good will among the peoples of the Campo region; regular Spanish–Gibraltarian contacts to increase mutual confidence; and the holding of a further seminar on the Gibraltar issue.[36]

There was to be one last attempt made to re-open the frontier between Gibraltar and the surrounding Campo region before the year ended when Señor Alejandro Rojas Marcos, the Leader of the Andalucian Socialist Party, tabled a question in parliament for the Spanish Government in which he asked about the possibility of the border being opened as 'a humanitarian gesture and evidence of good will towards the people of Gibraltar and that part of Andalucia which borders on it'. The lack of progress that had been made on the Gibraltar dispute was typified by the fact that the question could not be answered before parliament adjourned for Christmas.

8

SIGNS OF PROGRESS

Although the Gibraltar Government's responsibilities for the Rock's inhabitants were restricted to internal affairs, the general election of 1980, like the elections of 1972 and 1976, was contested around external questions; and one external question in particular dominated the campaigning, namely, that of Gibraltar's relationship with Spain.

The enforced isolation to which Gibraltar had been subjected throughout the 1970s, as a consequence of the frontier restrictions continually imposed by the Spanish Government and the resulting pre-occupation of the city's inhabitants with the Rock's future status, had left its impact upon the local political party system. The new Party for the Autonomony of Gibraltar (PAG), which had been founded by Mr Jose Emmanuel Triay and which had come into being after the last general election, had nominated three candidates for the 1980 campaign who were standing on a platform that called for a negotiated settlement with Spain; the ending of Gibraltar's colonial status and the development of a natural relationship with Spain.[1] Another new political party that had also come into being after the 1976 general election, the Democratic Party for British Gibraltar (DPBG) was a natural replacement for the former Integration With Britain Party (IWBP) which had ceased to exist immediately before that last election and, like the IWBP had previously done, the DPBG stood for the maintenance of close links between Gibraltar and Britain.[2] One party which had contested the last election had undergone a change of title: Mr Joe Bossano's Gibraltar Democratic Movement (GDM) had been renamed the Gibraltar Socialist Party (GSLP) in 1977 after three of its four elected members had crossed the floor of the House of Assembly.[3] The GSLP had essentially adopted the platform that had previously been assumed by the GDM; it continued to call for self-determination for Gibraltar and continued to advise against talks with Spain which it held had no claim to a say in the affairs of the Rock and its future. The GSLP fielded six candidates for the election most of whom, like Mr Bossano himself, were trade unionists but who, unlike him, lacked experience of politics. Sir Joshua Hassan's Association for the Advancement of Civil Rights–Gibraltar Labour Party (AACR–GLP) advocated its well-tried, well-tested and very successful formula of regular reminders to the British Government of Britain's guarantee that it would not hand over the people of Gibraltar to a foreign power against their freely and democratically expressed wishes.

The 1980 general election again drew forth a large number of candidates into the political fray but when the polling stations closed on 6 February, the turn-out was a relatively lowly 65 per cent; 10 per cent down on the 1976 figure and unusually low by the standards of Gibraltarian elections.[4] Although the number of candidates contesting the election had risen and the turn-out had gone down the result remained as it had done before with the eight candidates of the AACR–GLP all being returned and thus giving the party the 1-seat majority in the 15-seat House of Assembly it needed to retain office. The DPBG had made a spectacular electoral debut with 6 seats, thus ensuring that its Leader, Mr Peter Isola, would continue to serve as Leader of the Opposition. The GSLP paid the price for its inexperienced slate of candidates and also for the factionalisation which in 1977 had effectively destroyed its predecessor, the GDM, but the new Party's Leader, Mr Joe Bossano, had enjoyed a personal triumph in polling only 64 votes fewer than the veteran Chief Minister, Sir Joshua Hassan; his was, however, the party's sole success. Political continuity was thus assured by the election outcome and the depth of support for the continuation of past policy was reflected in the fact that the three candidates of the party that had sought to end that continuity through the development of a closer relationship with Spain, the PAG, had come bottom of the poll. It had become clear that the Rock's voters had again reaffirmed their faith in Sir Joshua Hassan's cautious approach to the problems of relations with Spain and had continued to endorse his insistence upon the voice of the city's inhabitants being heard in any Anglo-Spanish negotiations over Gibraltar's future status.

A resolute Gibraltar, attested by the outcome of the general election, and growing domestic pressure upon the Spanish Government to re-open the frontier both played a part in producing rare, unanimous support in the lower house of the Cortes for a resolution sponsored by the Spanish Socialist Party (PSOE) which called upon the Spanish Government to negotiate with Britain on the decolonialisation of Gibraltar and urged that the frontier with the Rock be opened 'whenever the progress of negotiations permits'. In a speech to the Congress of Deputies, after it had approved the PSOE's resolution, the Foreign Minister, Señor Oreja, said that 'if Great Britain accepts the UN mandate (to negotiate the future of Gibraltar) there will be a gradual return to normality in communications'. However, in the debate which had preceded the adoption of the resolution all of the political parties in the Congress had shown their continued support for the Spanish Government's claim to sovereignty over Gibraltar.[5] Nevertheless, the vote in the Congress had indicated that Madrid was now showing an increasing preparedness to find ways of breaking the impasse over Gibraltar and the first real sign of progress was to come very shortly when the British Foreign Secretary, Lord Carrington, and Mr Oreja were to discuss the Rock in Lisbon.[6]

Two days of Anglo-Spanish talks in Lisbon generated more progress on the divisive issue of Gibraltar than had been achieved in the previous eleven years of talks, talks about talks, UN resolutions and frontier restrictions. At the end of their meetings the two foreign ministers had produced what became known as the 'Lisbon Agreement'; an agreement that comprised six clauses:

1 The British and Spanish Governments desiring to strengthen their bilateral relations and thus to contribute to Western solidarity, intend, in accordance with the relevant United Nations resolutions, to resolve, in a spirit of friendship, the Gibraltar problem.
2 Both Governments have therefore agreed to start negotiations aimed at overcoming all the differences between them on Gibraltar.
3 Both Governments have reached agreement on the re-establishment of direct communications in the region. The Spanish Government has decided to suspend the application of measures at present in force. Both Governments have agreed that full co-operation should be on the basis of reciprocity and full equality of rights. They look forward to the further steps which will be taken on both sides which they believe will open the way to closer understanding between those directly concerned in the area.
4 To this end both Governments will be prepared to consider any proposals which the other may wish to make, recognising the need

to develop practical co-operation on a mutually beneficial basis.
5 The Spanish Government, in reaffirming its position on the re-establishment of the territorial integrity of Spain, restated its intention that in the coming negotiations the interests of the Gibraltarians should be fully safeguarded. For its part the British Government will fully maintain its commitment to honour the freely and democratically expressed wishes of the people of Gibraltar as set out in the preamble to the (1969) Gibraltar Constitution.
6 Officials of both sides will meet as soon as possible to prepare the necessary practical steps which will permit the implementation of the proposals agreed to above. It is envisaged that these preparations will be completed not later than 1 June.

When news of the Anglo-Spanish Lisbon Agreement reached Gibraltar it provided the basis for a joint statement by the Chief Minister and the Leader of the Opposition in which they expressed their qualified support.[7] The following day they departed for London where they had talks with Lord Carrington on matters relating to the Agreement.[8] After meeting with the Gibraltarian representatives Lord Carrington told the House of Lords that:

> it is envisaged that the practical preparations will be completed not later than June 1 which will then allow the Agreement to be speedily implemented. This is a very important step . . . though I should emphasise that this is only a beginning of what is likely to be a very long process.[9]

Two days later Lord Carrington's co-signatory to the Lisbon Agreement, Señor Oreja, told the Congress of Deputies that Spain was willing to recognise all the rights of the people of Gibraltar except their territorial sovereignty which 'does not and never has belonged to them'.[10]

The qualified support expressed by Sir Joshua and Mr Isola for the Lisbon Agreement stood in marked contrast to the response that was forthcoming from other quarters in Gibraltar when the details of the Agreement became known. On 13 April some fifteen hundred supporters of the Gibraltar Socialist Labour Party marched to the residence of the Governor of Gibraltar, General Sir William Jackson, and Mr Joe Bossano handed in a petition requesting that Lord Carrington exclude the 'question of Gibraltar's decolonisation and its future status' from any negotiations arising from the Lisbon

Agreement; in doing this the GSLP was remaining true to its manifesto pledge of opposing any negotiations on sovereignty with Spain because it held that Gibraltarian affairs were an improper area for Spanish involvement.

The GSLP demonstration came to symbolise what to outsiders appeared to be an enigmatic element in Gibraltarian politics, namely, the lack of enthusiasm for, and in some instances, outright opposition to, the Lisbon Agreement. Outsiders found it difficult to comprehend that a population which had endured eleven years of enforced isolation since the imposition of frontier restrictions by the Spanish Government would do anything other than welcome the prospect of a return to normality. Yet, by opposing the Lisbon Agreement, the Gibraltarians were effectively accepting a continuation of the hardships by which they had been so sorely afflicted; for unless the Agreement received their support it seemed highly improbable that Spain would lift its 'siege' of the Rock.

Gibraltarian opposition to the Lisbon Agreement was wide-ranging but may be summarised along the following lines:

1 The Agreement had been formulated and concluded over the head of the Government of Gibraltar in that there had not been any forewarning given to it of the nature of the meeting between Lord Carrington and Señor Oreja nor, more importantly, had there been any Gibraltarian representative invited to Lisbon; Sir Ian Gilmour, Lord Privy Seal, was to tell the House of Commons on 14 April that in all future negotiations the Gibraltarians themselves would be represented.

2 The Agreement went against the views of all the elected representatives in the House of Assembly in that Clause 4 stipulated that both the British and Spanish Governments were 'prepared to consider any proposals which the other may wish to make'; clearly this permitted a discussion on the sovereignty of Gibraltar. That this cause for concern was well-founded was revealed on 14 April in the House of Commons when Mr Peter Shore, Opposition spokesman on Foreign and Commonwealth Affairs, asked Sir Ian Gilmour, 'What subjects are to be covered in the future [Anglo-Spanish] negotiations or talks? Will he [Sir Ian] confirm that they will not include the question of sovereignty of Gibraltar?' Sir Ian Gilmour replied that, 'We agreed to talk about anything. That is part of the Agreement. Nothing is barred; we would discuss anything'.

3 In Clause 3 of the Agreement the Spanish Government was described as having only 'decided to suspend the measures at present in force', in consequence, many Gibraltarians feared that as this did not preclude the Spanish Government from re-introducing frontier restrictions – at some later stage they might again find themselves 'besieged'.

4 The suspicion existed in Gibraltar that the Spanish Government had been ready to sign the Agreement not out of any good will to the people of Gibraltar but because it needed the support of the British Government for Spain to become a member of the EC especially as French, and to a lesser extent, Italian opposition to Spanish membership was becoming increasingly vociferous.

5 In Clause 1 of the Agreement reference was made to 'relevant United Nations resolutions' but no reference was made to those specific resolutions which were to provide the parameters for future negotiations or talks on the Gibraltar question.

6 Clause 3 stipulated that 'future co-operation should be on the basis of reciprocity of full equality of rights' as a means of opening 'the way to clear understanding between those directly concerned in the area'. To many in Gibraltar this raised the threat that Spaniards would flock into Gibraltar, which enjoyed full employment, from the towns of Andalucia, where unemployment was high, and 'the unique character and culture of Gibraltar would be submerged'.[11]

7 The Agreement appeared to be founded on the assumption that there were only two possible scenarios for Gibraltar's future status; either to remain under British sovereignty or to pass into Spanish sovereignty. This apparently precluded the possibility of a third option, namely, independence for Gibraltar; an option which many in Gibraltar felt to be viable especially if it were to be linked to some form of economic union with either Britain or Spain or with both nations.

To these Gibraltarian grounds for opposition to the Lisbon Agreement were added two further anxieties which did not specifically arise from the Agreement itself but from the combination of the provisions of the Agreement with the forthcoming prospect of Spain becoming a member of the EC.

Firstly, it was surmised that with Spain a full member of the Community the restrictions on Spaniards entering Gibraltar would be ruled to be illegal and that following such a ruling Spaniards would then have the right to residence on the Rock and to establish businesses there; what the Spanish had failed to secure by 'siege' would in the future fall to them through purchasing power.[12] Secondly, fears were raised that once Spain entered the EC Gibraltar would find itself having to contribute monies to the common agricultural policy, assent to the common customs tariff and levy value added tax on goods and services; from all of which it was currently exempted.[13] If these latter measures were to come into force it was estimated that the Gibraltar Government would incur additional costs of £3.6 million per annum out of a total annual revenue of a little under £27 million.

It could be further questioned just what the Lisbon Agreement had actually achieved. It had certainly demonstrated the improved climate of Anglo-Spanish relations, Señor Oreja's 'new spirit', in that it showed a preparedness on the part of both Governments to try to resolve their differences over Gibraltar in a situation free from acrimony. It also demonstrated a new, and hitherto absent, Spanish willingness to offer concessions on the frontier restrictions which it had imposed upon Gibraltar in return for little more than a British pledge to engage in negotiations with an open agenda. Beyond this it left both London and Madrid continuing to specify their current positions on the question of Gibraltar's future status, for Clause 5 simply reiterated the Spanish Government's resolve to re-establish the territorial integrity of Spain and the British Government's determination to honour its pledge to act in accordance with the freely and democratically expressed wishes of the people of Gibraltar; a pledge already enshrined in the preamble to the Gibraltar Constitution of 1969.

However, the 'new spirit' which characterised Anglo-Spanish relations proved inadequate to surmount the difficulties occasioned by the re-statement of the fundamental objectives of both the British and Spanish Governments towards the question of the Rock's future status, and the frontier restrictions were not removed by 1 June.[14] In London, the Foreign and Commonwealth Office did not see the delay in the re-opening of the frontier as a serious set-back nor did it perceive its continuing closure as reflecting a lack of good will on the part of the Spanish authorities; instead the delay was portrayed as a question of timing which had been hampered by various technical difficulties encountered by the Spanish Government and which still had to be resolved in discussions between the British Embassy in Madrid and the Spanish Foreign Ministry. It fell to Señor Oreja to explain the passing of 1 June, the date set in the Lisbon Agreement for the lifting of frontier restrictions by the Spanish, with the border gates still closed.

This task he undertook in mid-June. Having emphasised his belief that the spirit of the Lisbon Agreement was still alive he did concede that he could not say when the frontier would be re-opened because of the need for the Spanish Government to organise administrative and physical facilities such as customs posts and parking areas. Although Señor Oreja denied that the delay was occasioned by any significant differences between the British and Spanish Governments he did suggest that more than technical difficulties had prevented the re-opening of the frontier when he stressed his Government's view that reciprocal treatment had to be granted to Spaniards on the Rock and Gibraltarians in Spain at the same time that the frontier gates were re-opened. What this statement implied was a disagreement between London and Madrid over the interpretation of Clause 3 of the Lisbon Agreement; the clause which referred to 'future co-operation . . . on the basis of reciprocity and full equality of rights'. Unless this phrase was clarified, and guarantees given, the Spanish Government feared that Spaniards would face discrimination in employment and residence in Gibraltar until Spain entered the EC.[15] A particular stumbling block was presented by the word 'future'. The Spanish Government suspected that the word would be used by the British Government to justify the British view that Spain should move first by restoring land communications between Gibraltar and mainland Spain before negotiations began; a perception which was markedly different to that of the Spanish Government which wanted the two events to coincide thus eradicating the prospect of discrimination against Spaniards seeking to move to the Rock and also as a way of guarding itself against criticism that it had made concessions to Britain while Britain had offered nothing in return.

In Gibraltar there was concurrence with the British interpretation that co-operation was tied to the removal of the frontier restrictions by the Spanish Government. It was further contended that the form that future co-operation could take should be determined by negotiation and that agreement would have to be reached during the course of that negotiation as to the matters to which reciprocity and full equality of rights was to be applied.

In the House of Commons, on 16 July, Sir Ian Gilmour was again asked by Mr Peter Shore about the situation in Gibraltar, 'has the timetable for negotiations for opening up communications between Gibraltrar and Spain been delayed? If so, why?' To which Sir Ian replied that, it has been delayed in the sense that he hoped the preliminary negotiations would have been completed by the beginning of June and that the frontier would have been open by now. 'We are still in discussion with the Spanish Government.' Differing interpretations of Clause 3 were, therefore, continuing to act as a real barrier to progress and were to continue to do so for, on 22 September, the Foreign and Commonwealth Office in London conceded that there had been some 'slippage' on the frontier agreement whilst emphasising that 'we certainly regard the necessary preparations on our side to have been completed'.

The Spanish Government was encountering difficulties on other fronts in addition to that posed to it by Clause 3 of the Lisbon Agreement. The target date for Spanish membership to the EC, provisionally set for 1 January 1983, was called into question by the French President, Giscard d'Estaing, when he argued that there should not be any further extension of the Community until such time as the continuing problems arising from the previous enlargement of 1973 had been satisfactorily resolved.[16] Although the French view was not supported by the other member nations of the Community, who re-affirmed their support for Spanish membership, the West German Federal Chancellor, Helmut Schmidt, stressed that 'without the indispensable adjustments to its agricultural policy and without a more balanced distribution of burdens the Community cannot finance the tasks which face its expansion southwards'.[17]

The response of the Spanish Government to this unenvisaged obstruction to its negotiations for membership of the EC was contained in a statement issued on 6 June in which it was noted that the 'new attitude' of the French President 'appears to be shifting difficulties which have arisen in other member countries on to the candidate countries'. The statement recorded that:

the [Spanish] Government wishes to make known its conviction that neither the internal measures taken by the Community to tackle its own problems nor its economic or other circumstances should be grounds for interrupting the course of negotiations or affecting the political commitment given so many times to Spain by the Governments of the Nine and especially by the President of the French Republic himself.

In the wake of this development intense diplomatic and political activity ensued between Spanish representatives and those of the Community member nations during the remainder of June and throughout July. On 3 July, the French Premier, Raymond Barre, visited Madrid and offered firm assurances to the Spanish of the French Government's continued support for Spanish accession whilst, at the same time, reiterating the necessity for a prior resolution of the Community's current budgetary and related issues.

Despite heated Spanish protests, French reservations appeared to have taken precedence and necessitated a re-thinking of the timetable for Spanish accession when a meeting of the Community's Council of Ministers, held in Brussels on 21-2 July, declined to endorse a Commission proposal that 1 January 1983 should be the official target date for formal Spanish membership of the EC. The Council of Ministers, however, pledged itself to 'uninterrupted negotiations' to bring Spain into the Community 'as soon as possible'.[18]

The difficulties confronting the Spanish Government in its pursuit of membership of the EC occasioned concern to those who had perceived a solution to the Gibraltar problem lying in Spain's full involvement in the wider context of Europe. At the same time, those same difficulties gave encouragement to others who felt that the British Government had singularly failed in the past to use the threat of withdrawal of its support for Spanish membership of the Community as a lever with which to wrench concessions from Spain. With the date stipulated in the Lisbon Agreement for the completion of negotiations leading to the re-opening of the frontier having passed with the frontier restrictions still in place, it was this body of opinion which looked to Britain to employ what it perceived to be a weighty sanction to oblige Spain to comply with Clause 3 of the Agreement and particularly that part of it which stated that 'the Spanish Government has decided to suspend the application of the measures at present in force'.

Therefore, it was with great interest that a meeting between Lord Carrington and the new Spanish Foreign Minister, Señor Jose Pedro Perez Llorca, was awaited. That meeting took place in New York in September but, apart from mutual assurances that both signatories to the Lisbon Agreement continued to see themselves committed to it, no progress was made and the Gibraltar question again appeared to have been consigned to limbo.

As the year ended Clause 3 once again rose to prominence and became the subject of discussion not in London nor in Madrid but in Gibraltar. On 30 December, the House of Assembly adopted a resolution that Spaniards should not enjoy rights equal to those of citizens of European Community countries in Gibraltar until Spain became a full member of the Community. With this final fling it seemed that 1980 had indeed been the year of Clause 3 and that 1981 offered few prospects for rapid change.

9

OTHER PRIORITIES

Anglo-Spanish talks on Gibraltar did not begin favourably in 1981. In January, Sir Ian Gilmour, the Lord Privy Seal, visited Madrid for talks with Spanish Foreign Minister, Señor Perez Llorca. The talks were stated by Sir Ian not to involve negotiations over Gibraltar's future but were to concentrate on two other areas of Spanish foreign policy, namely, the difficulties standing in the way of Spanish negotiations to join the EC in the near future and proposals relating to Spain's possible membership of NATO. However, even if by now overshadowed by questions of Community politics, Gibraltar was discussed; if for no other reason than because of Spanish displeasure at the Gibraltar House of Assembly resolution of 30 December 1980. Señor Perez Llorca demanded equality for Spaniards going to the Rock, in anticipation of their rights after Spain became a full member of the EC, as a condition for reopening the frontier with Gibraltar. He called upon the British Government to induce the Gibraltarian authorities to make such changes as were necessary to achieve this for otherwise, he argued, if Spain was to unconditionally reopen the frontier its citizens would find themselves disadvantaged on the Rock. Sir Ian's response was to claim that it was a 'myth' that Spaniards would face discrimination in Gibraltar if Spain was to reopen the frontier and assured the Spanish Government that Britain was remaining true to the Lisbon Agreement and that negotiations on Gibraltar's future would begin simultaneously with the lifting of Spanish restrictions on the Rock. Questioned as to whether the failure to obtain a rapprochement constituted a backward step, Sir Ian maintained that there had 'been no going back, at most we have not gone forward'. In short, disagreements arising from Clause 3 seemed to have rendered the Lisbon Agreement unworkable and left the impasse in Anglo-Spanish relations on the question of Gibraltar's future as deep as ever.

Matters other than its claim to the sovereignty of Gibraltar now began to occupy the Spanish Government. The fragility of Spanish democracy had been revealed by the attempted coup of February but the coup's lack of success had ensured that civil power was more firmly established in Spain and that the armed forces were no longer the final arbiters of policy.[1] The EC had viewed the applications for membership from Greece, Portugal and Spain as arising from a common political motive; to succour their young democratic institutions and processes. It now seemed that with the Spanish application to join the Community likely to be a long drawn out process, because of French objections, and with the recent challenge to Spanish democracy having raised questions about the stability of the new regime, that Spanish membership of NATO was viewed as affording a vehicle through which democracy could be strengthened. The Spanish Prime Minister, Leopoldo Calvo Sotelo, indicated that Spain's membership of NATO could be achieved by the end of the year. NATO had been absent from Madrid's foreign policy agenda for two years but now Señor Sotelo described membership of the Organisation as the 'natural culmination of [Spain's] option for Europe' and promised to conduct talks with all of the political parties in the Cortes before a final application was made.[2]

Despite the divisions which existed over Gibraltar's future status the British Government had continued to lend its support to the Spanish Government's application to join the EC. The question now being posed was, would Britain do the same in respect of the Spanish application to join NATO given that the Lisbon Agreement had not led to Spain ending the frontier restrictions which it had imposed upon the Rock? It was with the intentiom of finding means by which the impasse on Gibraltar could be broken that five members of the House of Commons Select Committee on Foreign Affairs visited Madrid and Gibraltar in late April.[3] In Madrid, Señor Ignacio Camunas, the Senate Foreign Affairs Committee chairman, indicated that his Government's decision to apply for membership of NATO would permit a solution to the question of Gibraltar's future status and the problem posed by the Rock's usage as a military base.

The House of Commons Select Committee's report of its findings was published on 28 August. It concluded that 'there is clearly no solution to a problem which has vexed relations between the UK and Spain for more than two centuries' but it emphasised that 'we believe that for a number of reasons the time is now ripe for some movement'. Having recognised that when Spain became a member of the EC the continuation of frontier restrictions on Gibraltar would be inconceivable, the Report tackled the problematical Clause 3 of the Lisbon Agreement and recommended that:

1 as on the day on which Spain joined the Community the status of Spaniards in Gibraltar would change in compliance with EC regulations, it would only anticipate this event by a short time if the British Government was, 'on the lifting of [frontier] restrictions, to give to Spaniards in Gibraltar the rights that they would have as citizens of an EC country';
2 any laws or regulations enacted to guarantee Spaniards in Gibraltar the rights enjoyed by Gibraltarians in Spain should comply with the 'reciprocity and full equality of rights' stipulated in Clause 3 of the Lisbon Agreement; and that
3 on the day that Spain ended the frontier restrictions the British Government should give a formal assurance that negotiations to resolve all differences on Gibraltar would begin.

Additionally, the Report recognised that the continued use of Gibraltar as a military base by Britain would depend upon Spanish good will and that Gibraltar could not at present survive without the naval dockyard and other military installations. A possible resolution to the dispute was seen as lying in the regional autonomy stipulated in the Spanish Constitution. The Report also addressed the difficult question of the sovereignty of Gibraltar; whilst it recognised that the British Government could transfer its sovereignty over Gibraltar to another power, it equally recognised that it could not do so against the democratically and freely expressed wishes of the Gibraltarians but it emphasised that their veto 'should not be allowed to prevent the exploration of possible constitutional solutions seeking to reconcile the conflicting interests of all parties'.

The Select Committee's report set alarm bells ringing in Gibraltar when it stated that although the British Government clearly had obligations to Gibraltar it 'might well find itself in grave difficulties if it does not make clear to the Gibraltarians that, though the Colony is in many respects unique, the British Government's primary responsibility is to [the British] Parliament'. This was by far the most radical statement of British perceptions towards the relationship between Britain and Gibraltar in that it could be taken to imply that what had been given by way of a constitution could also readily be taken away through constitutional change and that the constitution granted to the Rock in 1969 which had, in its preamble, afforded the Gibraltarians the power of a veto over their future was not inviolate. Furthermore, it emphasised this point by observing the minor status of the British Government's responsibility to the people of Gibraltar in relation to its wider responsibilities to those who were directly represented in Westminister.

Before the Select Committee's report was published Lord Carrington and Señor Perez Llorca met in Brussels. There was speculation in Madrid that the Spanish Government was willing to renounce its demand for equal treatment for Spaniards in Gibraltar as a prior condition for reopening the frontier and that it was also prepared to delay any further discussion on the question of Gibraltar's sovereignty.[4] There was also contrary speculation that no new solutions to the Gibraltar problem would be offered to the British by Señor Perez Llorca.[5] In the event, Señor Perez Llorca's visit to Brussels was to prove fruitless. Not only did the French reject the Spanish argument that Spanish entry negotiations should be held concurrently with the Community's internal reforms to its budgetary processes but neither was progress made in his meeting with Lord Carrington.[6] Again Señor Perez Llorca said that Madrid was not demanding that Spaniards on the Rock be given equality of rights with Gibraltarians but, as a first step, the same rights as were granted to EC citizens in Gibraltar.[7] Lord Carrington offered the traditional reply to this request, namely, that such equality could only be tendered when Spain became a full member of the Community. Clearly, no new Spanish proposals had been forthcoming. It thus appeared that unless some fresh approach to the Gibraltar question was found the Lisbon Agreement, which had been so readily arrived at, would continue to be unworkable in practice and that the frontier would remain closed until 1984 – the earliest possible date for Spanish entry into the Community.

Lord Carrington and Señor Perez Llorca met again in Madrid on 16 August.[8] Although the subjects of Spanish accession to the EC and NATO were again discussed a Spanish Foreign Ministry spokesman was reported as claiming that it was on the topic of Gibraltar that the greatest progress had been made and that there were now strong grounds for believing that a solution was in the offing. Such news produced further speculation in Gibraltar that the frontier would soon be reopened but it did not generate an official response from the Gibraltar Government nor did it elicit a warm welcome from the Rock's inhabitants. It seemed that Gibraltarians were content to opt for the relative security that they had enjoyed since the frontier was closed. There was a widespread feeling that a democratic Spain was no different to Franco's Spain when it came to good will towards the city.

The Spanish Government, however, was confronted by problems other than Gibraltar. From May through August five rounds of talks were held between Spain and the United States on the renewal of the 1976 Treaty on US bases on Spanish territory.[9] On 3 September both nations agreed to suspend negotiations and extend the Treaty for a further eight months because of complications which had arisen from Spain's pending application to join NATO and because of the Spanish Government's demand for more favourable terms from the US than it had received in 1976.[10] In particular, the Spanish Government contended that, as the 1976 Treaty had been negotiated during the last months of the Franco regime from what the present Spanish Government now considered to be a position of weakness, new terms and conditions would have to be formulated and agreed. In negotiating a new treaty, Spain was seeking a defence guarantee and a greater financial commitment from the US as well as modern military equipment and weaponry together with the transference of US technology to Spain for use by the Spanish armaments industry; this, the Reagan Administration was not initially prepared to concede.

By late summer 1981, Señor Perez Llorca had completed soundings of current NATO members on Spanish entry to the Organisation. Most of the member countries had revealed themselves to be in favour of Spanish membership although concern had been expressed by some that such an extension to NATO would disturb the East–West balance of power and lead to the Warsaw Pact strengthening its military capacity.[11] The Soviet response was, however, muted. On 7 September, a Soviet Note had been passed to Señor Perez Llorca by the Soviet Chargé d'Affaires, Mr Igor Ivanov, which warned of the 'negative consequences' for Spain if it were to join NATO. The Note emphasised that whilst the Soviet Union 'has always respected . . . Spain's sovereignty' it could not ignore 'plans or activities which may lead to the escalation of international tension and do great harm to the cause of European security'. Señor Perez Llorca returned the Soviet Note and claimed that it constituted 'unacceptable interference' in Spanish affairs.[12]

The Spanish Government, however, was less troubled by Soviet opposition to its prospective membership of NATO than it was by domestic opposition. Appreciating that all Spanish political parties were united on the Spanish claim to sovereignty over Gibraltar, Señor Perez Llorca argued that membership of NATO would strengthen Spain's claim to Gibraltar whilst Defence Minister, Señor Alberto Oliart, maintained that Spanish membership offered the only hope of resolving the Gibraltar dispute.[13] Government and Opposition, however, held differing views on the effect that Spanish membership of NATO would have on the question of Gibraltar. Señor Felipe Gonzalez, the opposition Socialist Party leader, argued that, with Spain in NATO, an attack on Gibraltar would oblige Spain to commit its forces in defence of the Rock and contended that if Spain did indeed enter NATO the Spanish Government should demand that other NATO countries recognise Spanish sovereignty over Gibraltar. On 31 August the Government sent its petition for parliamentary approval for Spain to join NATO to the Congress of Deputies and debate opened in the Foreign Affairs Committee on 6 October.

Despite a plethora of amendments, on 8 October, the Foreign Affairs Committee authorised the Government to seek accession to NATO.[14] The Committee's final resolution recommended that the Government should give special consideration to the following when conducting the negotiations that would take Spain into the Organisation:

1 that in view of the threats to international detente it was now more than ever necessary to obtain defence guarantees for Spain;
2 that it was necessary to guarantee the security of all the national territory, both mainland and extrapeninsular;
3 that 'the recovery of Spanish sovereignty over Gibraltar is fundamental, as is the strengthening of defence and sovereignty over the whole of Spain';
4 that parallel with the NATO negotiations, negotiations on Spanish entry to the EC should also be accelerated.

The debate on Spanish accession to NATO opened in the Congress on 26 October and the application was approved on 29 October by 180 votes to 146.[15] During the course of the debate attention was directed to the third of the Foreign Affairs Committee's recommendations; that concerned with the Spanish claim to sovereignty over Gibraltar. Speaking in the debate, Señor

Calvo Sotelo said that Spain's neutrality might have weakened its claim and assured the Cortes that 'the Government is determined to make progress on this claim and has good reason to believe that such progress will be assured if Spain signs the Treaty of Washington'.[16] For the opposition Socialist Party its leader, Señor Gonzalez, said that it was a mistake to think that Spain must give up its claim to Gibraltar if it did not enter NATO. When the Government's petition received final approval in the Senate on 26 November, by 106 votes to 60 with one abstention, Gibraltar was again raised.[17] Speaking for the Socialist Party, Señor Fernando Moran, called on the Government to obtain formal recognition of Spanish sovereignty over Gibraltar before joining NATO. The Government, however, responded that it would be more realistic and pragmatic to raise the question of Gibraltar's future status after Spanish accession.[18]

Spain finally applied to join NATO on 2 December and, at a ceremonial plenary session of the North Atlantic Council in Brussels on 15 December, NATO Foreign Ministers signed a Protocol of Accession inviting Spain to become the sixteenth member of the Organisation.

Whilst the debate on membership of NATO had dominated Spanish political life and whilst reference to Gibraltar had frequently arisen in the course of that debate, the Gibraltarians were pre-occupied with questions more fundamental and of greater immediate import than whether a Spain within or without NATO would advance Spanish claims to sovereignty over the Rock. In June 1981, the Secretary of State for Defence, Mr John Nott, presented a Defence White Paper to the British Parliament; the culmination of an extensive review of Britain's defence requirements. The White Paper stipulated that a reduction was to be made in the number of surface warships to be retained by the Royal Navy. In Gibraltar this caused understandable concern for the Gibraltarian economy was heavily dependent on the Rock continuing to serve as a British naval base and upon the retention of the naval dockyard. The full implication of the White Paper became evident on 23 November when, in a written parliamentary reply, Mr Nott stated that, 'changed plans for the Royal Navy no longer sustain a need for a naval dockyard in Gibraltar' and that, following such preparatory action as was deemed necessary in 1982, the closure of the dockyard would commence in 1983. Mr Nott promised that consultations would begin with the Gibraltar Government and local trade unions to consider possible economic alternatives among which would be the commercialisation of the dockyard. Nor was the naval dockyard the only facility in Gibraltar to be endangered as a result of the defence review, for Mr Nott noted that it was planned to hold discussions on reducing the opening hours of the RAF airfield at Gibraltar, which also catered for civilian air traffic, to 'bring them more into line with those required for military purposes'. It was emphasised by the Ministry of Defence in London that the naval base and refuelling facilities would not be closed and that, therefore, Gibraltar would still retain its key naval role in British military planning.

There had long been speculation about the future of the naval dockyard in Gibraltar as its importance had lessened with the withdrawal of a permanent Royal Navy presence from the Mediterranean.[19] That the decision to close the dockyard had not come as a complete surprise to Gibraltar was attested to by the Rock's Chief Minister, Sir Joshua Hassan, who observed that 'We've been prepared for it and done a lot of homework. With help from the British Government we hope to save the bulk of the dockyard workforce.'[20] The exact details of how 'the bulk of the dockyard workforce' was to be saved were not revealed. Initial estimates suggested that 950 jobs would be lost in the dockyard during the first phase of its closure, of which some 800 would be those of local people, and that eventually 3,000 jobs would disappear of which half were currently held by Gibraltarians.[21] In short, as the dockyard was the single largest employer on the Rock, its closure would mean the loss of 15 per cent of all jobs and this would have a knock-on effect throughout the local economy.

Sir Joshua was to concede that he was more perturbed by the announcement that the airfield was to operate for fewer hours each day. He described the airfield as Gibraltar's lifeline and stressed that 'This is something we will have to resist most strongly. I don't think this is going to be workable. The airfield facilities must be available for civilian flights.'[22] As a growth in tourism had been identified as being a central element in a revitalised economy the reduced use of the airfield whilst the Spanish frontier restrictions were still in force implied that this option was unlikely to remain open at the very time when the need for it was most pressing.

A team of British officials flew to Gibraltar on 24 November, the day following Mr Nott's announcement of the dockyard closure, for talks with Sir Joshua and other members of the Government.[23] Sir Joshua himself visited London on 14 December where he had a meeting with Lord Carrington. After this latter meeting it was announced that the British Government would offer an immediate increase in aid to Gibraltar pending the final negotiations on the closure of the dockyard. It was, however, also announced that the closure decision was irreversible. When, having returned to Gibraltar, Sir Joshua told the House of Assembly of the irrevocable nature of the British Government's decision a demonstration took place outside the House.

With Spain well advanced down the road to membership of NATO and progressing more slowly towards membership of the EC, the British Government's decision led many Gibraltarians to draw the conclusion that the guarantee contained in the preamble to their Constitution might be more apparent than real. While the British Government was pledged not to transfer sovereignty over the Rock to another power against their democratically and freely expressed wishes it might either surrender that sovereignty, in which case the Rock would return to Spain, or simply render the city economically non-viable by withdrawal of financial investment.

To these concerns were to be added anxieties over the implications for Gibraltar of the reopening of the frontier by the Spanish authorities; anxieties which were to be brought into sharp focus following talks between the British Prime Minister, Margaret Thatcher, and her Spanish counterpart in London on 8 January 1982. Although the talks covered Spain's proposed membership of both NATO and the EC and although British support was again forthcoming for both of those ventures, it was on the subject of Gibraltar that significant progress appeared to have been made. A joint communiqué, issued at the end of the meeting between the two premiers, stated that 'Both countries have agreed to commence, on 20 April 1982, the negotiations envisaged in the Lisbon statement with the aim of overcoming the differences between them on Gibraltar. On the same day direct communications will be re-established as provided for in the Lisbon statement.'[24] At a following press conference, Señor Calvo Sotelo contended that 'the definitive solution of the [Gibraltar] problem must be the re-establishment of the territorial integrity of Spain' but he stressed that 'this objective is consistent with the intention expressed by the Spanish Government to the effect that the interests of the Gibraltarians shall be safeguarded'. The stumbling block to Anglo-Spanish negotiations on Gibraltar, Clause 3 of the Lisbon Agreement, appeared to have been resolved by the leaders of the two nations. Señor Calvo Sotelo said that he had received assurances from the British Government that the talks scheduled for 20 April were intended to 'resolve all differences over Gibraltar on the very day that the barrier is lifted' and that all residential, social security and other 'discriminations' against Spaniards in Gibraltar would be removed. The impact that Gibraltar had exerted upon Anglo-Spanish relations was also recognised by the Spanish Premier who mentioned that 'the only problem, although a grave one, which separates Spain and the United Kingdom has now entered negotiation stage. Relations between the two countries must likewise enter a new and promising stage'.

With European and Atlantic aspirations serving to draw London and Madrid closer together many in Gibraltar wondered how long the divisive issue of the Rock's sovereignty would be permitted to sour the growth of a rapprochement born out of the identification of common interests.

On 4 February it was the turn of Señor Perez-Llorca to tell the Foreign Affairs Committee of the Congress that the Spanish Government still regarded Gibraltar as an integral part of the national territory. As though to emphasise that progress was at last being made on the question of the Rock, he announced the creation of the post of Under-Governor of the Campo and the establishment of two working parties to ensure that the re-opening of the frontier would not be disruptive to the region nor produce an imbalance between its towns and Gibraltar.[25] This decision also reflected the concern of the Spanish authorities that when the frontier was reopened Spaniards would travel to the Rock to purchase goods which were subject to high duties in Spain. The reopening of the frontier was also the principal item on the agenda when Señor Perez-Llorca and Lord Carrington met in Brussels on 23 March.

However, more distant events were now to exert an influence on the Gibraltar problem. On 2 April Argentina launched a military invasion of the Falkland Islands.[26] In Gibraltar there were fears that similar action might be taken by Spain to recover the Rock. Certainly, Spain's extreme right had been vociferous in its acclamation of Argentina's direct use of force in the South Atlantic as a means of resolving a long-standing and intractable political problem. More moderate political opinion in Spain expressed satisfaction at Argentina's objectives in the Falklands but was critical of the methods that had been employed; Spain's own recent history of military influence and interference in the domestic political process had made many wary of the use of armed force as a vehicle for the attainment of political ends. For example, the Spanish

Socialist Workers' Party, the major opposition party, had expressed disapproval of British 'colonialism' and had afforded recognition to what it had described as Argentina's legitimate rights over the Falkland Islands but, at the same time, had condemned the use of force and recommended that the problem be resolved in accordance with UN principles.

Nevertheless, the close proximity of Spain's scheduled ending of the blockade of Gibraltar and the Argentinian seizure of the Falkland Islands led many in Spain to link the two events. Such a linkage was not made by the Spanish Council of Ministers which, on 3 April, issued a statement containing an assurance that the Spanish Government regarded Gibraltar and the Falklands as separate questions. Similarly, whilst denying that there was an exact parallel between the two cases, Señor Calvo Sotelo felt that Spain would not have to wait long before recovering sovereignty over the Rock.[27] Yet others foresaw the Falkland crisis as having disadvantageous consequences for Spain's claim to the recovery of its sovereignty over Gibraltar. They contended that there was likely to be a hardening of the resolve of the Gibraltarians to maintain the status quo and that British politicians and diplomats were likely to come under increasing pressure, given the outpourings of nationalist sentiment occasioned in Britain by the Argentinian invasion, not to give ground on the Rock.[28]

The determination of the Spanish Government that events in the South Atlantic should not impact upon events in the Straits of Gibraltar was evident at the UN. When presented with Security Council Resolution 502, which called for the cessation of hostilities between Argentina and Britain and a withdrawal of Argentine military forces from the Falkland Islands, Spain abstained; a vote which marked something of a retreat from the consistent support it had previously afforded Argentina's claim to sovereignty over the Falklands in the UN General Assembly.

On 6 April, the Spanish Diplomatic Information Office went so far as to state that neither Argentina's invasion of the Falkland Islands nor the resignation of the British Foreign Secretary, in the wake of that event, would alter Spain's plans to implement the Lisbon Agreement.[29] Yet three days later statements were issued in both Madrid and London announcing that the Anglo-Spanish talks scheduled for 20 April had been postponed until 25 June. Unquestionably, the preoccupation of the British Government with the Falkland crisis was the cause of this decision. Nevertheless, there was confirmation that both the British and Spanish Governments remained committed to the Lisbon Agreement, to reopening direct communication between the Rock and the adjoining Spanish mainland and to developing closer understanding among those directly concerned in the area of Gibraltar. The negotiations re-scheduled for 25 June had the purpose of 'overcoming all the differences between the two sides on Gibraltar'.

Two days before the date set for Spanish accession to NATO the Spanish Socialist Workers' Party (PSOE) tabled a motion in the Cortes calling on the Government to postpone Spain's entry until the British Government had given 'full assurances' that it would restore the sovereignty of Gibraltar to the Spanish people.[30] The timing of this PSOE motion, coming when it was already too late to prevent Spain entering NATO, was rejected by the Spanish Government as a propaganda move and Spain became the sixteenth member of the Organisation on 30 May 1982 after the Spanish Chargé d'Affaires in Washington, Señor Alonso Alvarez de Toledo, deposited the formal instrument of ratification with the US Deputy Secretary of State, Mr Walter Stoessel.[31]

With the next round of Anglo-Spanish talks on Gibraltar scheduled to take place in a little over three weeks, there was speculation in both Spain and Britain as to the effect that the presence of both nations in NATO would have upon the meeting. In Madrid, it was felt that Spain would seek the right to use the naval base at Gibraltar together with its associated facilities. In London, there was a disinclination to welcome the Spanish fleet to the Rock until such time as the negotiations on the reopening of the frontier and the implementation of the Lisbon Agreement had made real progress. It was further observed that the naval base at Gibraltar could not operate effectively and efficiently in NATO's interest until proper land, sea and air communications were restored by the Spanish authorities. The fact that there had not been any debate as yet on the exact nature of the military role that Spain would play in NATO once it had been integrated into the military command structure added a further element of uncertainty, although some Spanish politicians envisaged a new command structure for the Organisation with Spanish control of the strategic arc which extended from the Balearic Islands through Gibraltar to the Canaries.

In the event, such speculation was neither proved nor disproved because when the British and Spanish Foreign Ministers met in Luxembourg, on 21 June, at a meeting of the EC to continue negotiations on Spanish entry to the Community, it was announced that both the Anglo-Spanish talks and the reopening of the frontier which had been postponed from April had been indefinitely postponed. In a joint communiqué it was stated that the talks set for June had been postponed 'at the suggestion of the Spanish Government' but it was stipulated that both nations were 'determined to keep alive the process opened by the Lisbon Agreement'. Furthermore, both Governments agreed to maintain contact 'personally and through diplomatic channels' and to determine a new date for a meeting in due course. The Spanish view was that the scheduled talks would have been unlikely to have succeeded as British attitudes on the question of Gibraltar had hardened as a consequence of the Falkland crisis. Additionally, it was unofficially suggested that Señor Perez-Llorca had indicated to Mr Pym that some concession should be made by the British Government on the question of the sovereignty of Gibraltar as a means of countering the anti-British sentiment that the conflict in the South Atlantic had produced in Spain. Whilst denying that the sovereignty of Gibraltar had been raised during his meeting with Señor Perez-Llorca, Mr Pym admitted that there was a real danger that had the talks begun they would not have been fruitful and, had any breakdown occurred, that would have made matters worse than they already were. Mr Pym emphasised that Spanish enthusiasm to join the EC would, in his opinion, be the key factor in securing the ending of the Spanish authorities' frontier restrictions on Gibraltar; but given that negotiations on Spanish accession to the Community were making little headway, it seemed that Gibraltar would remain 'besieged' for the immediate future.[32]

In Gibraltar itself, the feeling was that the British Government should exhibit some of the resolve it had evidenced in dealing with Argentina over the Falkland Islands; that the Lisbon Agreement should be dropped; and Spain required to remove the frontier restrictions without delay, now that it was both a member of the Council of Europe and of NATO, and respect the right of people to move freely. Few on the Rock held out any hope that the Anglo-Spanish talks would recommence or that the frontier would reopen until after the next Spanish general election had been held in the autumn. Even so, restrictions at the frontier were eased from July onwards after an announcement by the Governor of Cadiz Province, Señor Jose Gonzalez Palacios, that more flexibility would be shown by Spain in granting permission to cross the border for humanitarian reasons. The Under-Governor of the Campo, Señor

Salvador Camino Crespo, noted that this flexibility should not be taken to imply any change in the positions assumed by either the Spanish or British Government on the question of Gibraltar's sovereignty.

When the Spanish electorate went to the polls on 28 October they returned a left-wing government to office in Madrid for the first time in more than forty years. The Spanish Socialist Workers' Party had promised that, if returned to power, it would reopen the frontier with Gibraltar. That pledge had not been born solely out of humanitarian reasons but owed its origin also to the fact that the region from which it drew its greatest electoral support, Andalucia, had been the area of Spain which had experienced by far the greatest economic hardship as a consequence of the frontier restrictions.[33] In the Cortes on 30 November, the Premier elect, Señor Felipe Gonzalez Marquez, promised that there would be immediate negotiations on both the lifting of the frontier restrictions and on the integration of Gibraltar into Spanish territory.[34] Having specified 14–15 December as the date upon which the frontier between the Spanish mainland and Gibraltar would be reopened, Señor Gonzalez said that this was being done for 'humanitarian reasons' and not as a part of the process envisaged in the Lisbon Agreement; consequently, there had not been any prior consultation with the British Government although he envisaged that this action 'would aid the climate for negotiation'. However, this was only to be a limited reopening of the frontier since, to prevent possible 'negative economic' effects being experienced by the towns of the Campo region, the Spanish Government declared that individuals would only be allowed to cross the frontier once per day.[35]

At a NATO meeting held in Brussels on 9–10 December, the newly appointed Spanish Foreign Minister, Señor Don Moran Lopez, had discussions with Mr Pym on the question of Gibraltar's future. It was agreed by the two foreign ministers that negotiations covering the technical details of the full reopening of the frontier should be held in early 1983.

The partial reopening of the frontier and the prospect of Anglo-Spanish negotiations on its full reopening met with a mixed response in Gibraltar where many were of the opinion that strict Spanish customs controls would serve to nullify any significant increase in trade for the Rock, irrespective of whether the frontier was partially or fully open. Whilst Sir Joshua Hassan welcomed the limited reopening of the frontier as 'a step in the right direction', the restrictions imposed on access announced by the Spanish Government on 11 December caused the Gibraltar Government to declare, two days later, that it would close its side of the border between 1.00 am and 6.00 am daily.[36] This was the first time since the frontier restrictions were imposed by the Spanish Government in 1969 that the Gibraltarian authorities had closed the gates on their side of the border and raised the ironical question as to who was besieging whom? In practice, the Gibraltar Government did not implement its decision because of the speedy intervention of Mr Pym and it was rescinded the day after it was made public. The decision of the Spanish Government raised immediate concern among the three thousand Moroccan day-workers in Gibraltar who had originally been brought into the Rock to replace the Spanish workers who had been prevented from entering the City since 1969. The Moroccans foresaw a real threat to their continued employment when faced by competition from the large number of unemployed workers in the Campo region.

10

HOPE RENEWED

It appeared to have become mandatory that in the first few weeks of each new year a statement would emanate from Madrid reiterating the Spanish claim to sovereignty over Gibraltar. The year 1983 was not to prove to be an exception to that pattern when on 26 January the Spanish Foreign Minister, Señor Fernando Moran, ruled out any further Anglo-Spanish talks on the future of Gibraltar unless the question of the Rock's sovereignty was on the agenda. He boldly asserted that 'no Spanish foreign minister can sit down to negotiate if sovereignty is not discussed' and by so doing he immediately threw into doubt the talks which had provisionally been set for the spring.[1] In London, the response to Señor Moran's declaration was decidedly low key with the Foreign and Commonwealth Office citing the Lisbon Agreement as affording both the British and Spanish Governments the opportunity to raise any proposals they wished, including the question of Gibraltar's future sovereignty. Gibraltarians were troubled by the preparedness of the British Government to enter into talks with Madrid on the question of the Rock's sovereignty, just as they had been since the signing of the Lisbon Agreement, and they drew a parallel between their own situation and that of the Falkland Islands. In the case of the Falklands the British Government had refused to negotiate with the Argentine Government and had claimed that there was no scope for talks aimed at resolving their disagreements because any talks that did take place would be used by the Argentinians as a means of achieving a direct transference of sovereignty over the Islands from Britain to Argentina against the expressed wishes of the islanders.[2] In short, viewed from Gibraltar, there appeared to be a glaring inconsistency in the attitude of the British Government between the stance it had adopted on two essentially similar issues; an inconsistency adversely weighted against the Rock's inhabitants.

Gibraltarians were not only concerned by the nature of any further agreements that might emanate from future Anglo-Spanish negotiations, for they also had the prospect of the impending closure of the naval dockyard with which to contend and here support came from an unexpected source. During a visit to London, Admiral William Crowe, Commander in Chief US Naval Forces in Europe, raised doubts as to the wisdom of the British Government's decision to close the dockyard. Having noted that the yard was one of the few which possessed the capability to handle nuclear-powered submarines, a facility the US Navy had not to date had cause to use, he emphasised the continuing strategic importance of Gibraltar to the defence of the Mediterranean region. In particular, he pointed out that, in the event of war, one of the West's first tasks would be to isolate the Mediterranean Sea from the Atlantic Ocean and expressed the opinion that NATO had so far failed to appreciate the strategic importance of the Mediterranean, and by implication that of Gibraltar, to its southern flank. Admiral Crowe's views were supported later in the year by assessments of Gibraltar's strategic importance which emanated from NATO. It was contended that whereas the eastern Mediterranean was well served by the US Sixth Fleet and the concentration there of Italian, Greek and Turkish forces, the western Mediterranean was only sparsely populated by surface naval craft. More especially, there was concern that Britain was failing to make adequate usage of the defensive potential of Gibraltar by not remedying the lack of surveillance radar, missiles and guns and by stationing an inadequate air capability to command the strategically important straits.[3] The prospect of the privatisation of the naval dockyard raised further doubts as to whether, under civilian ownership, it would be able to continue to provide maintenance facilities for hunter–killer, nuclear submarines. With Spanish forces not yet integrated into NATO's military command structure these factors were seen as leaving the Organisation's southern flank dangerously over-exposed to any concerted military challenge whilst Gibraltar was perceived as offering the means by which this deficiency could be overcome with minimal expenditure.

The impending closure of the naval dockyard and its transference to private ownership not only had implications for NATO's military strategy in the Mediterranean but also carried adverse connotations for the Gibraltarian economy. This latter matter was raised in the House of Commons where it was noted that Gibraltar's future was dependent upon its ability to remain economically viable. The Foreign Secretary, Mr Francis Pym, contended that the implementation of the Lisbon Agreement would be of economic benefit to Gibraltar but that irrespective of the Agreement's implementation the Overseas Development Agency had set aside £13 million for the period 1981 to 1986 to facilitate a development programme for the Rock.[4]

The Gibraltarian economy was certainly in need of financial support. The partial frontier restrictions enforced by the Spanish authorities from 1964 to 1969 and the total closure of the border from that latter date had wrought havoc upon the local economy. Those who had anticipated that commerce on the Rock would improve with the partial reopening of the frontier were confounded by the realisation that far from proving to be advantageous to trade the reverse had occurred. With the Spanish customs stopping the movement of goods from Gibraltar to the Spanish mainland but not from Spain into Gibraltar the Spanish border towns of La Linea, San Roque and Algeciras were enjoying a mini economic boom whilst Gibraltar was experiencing a cash outflow to Spain of between £100,000 to £150,000 per week. The restrictions imposed upon free movement across the frontier prevented people moving from Gibraltar into Spain and meant that the Rock's tourist industry, a sector marked out for future growth, could not yet be revived. The pending closure of the naval dockyard and the ensuing gloomy prospects for employment added both to the semblance and reality of an economic crisis. Additionally, there were complaints as to the way in which the Ministry of Overseas Development authorised the spending of money under the 'support and sustain' undertaking given to the Gibraltar Government by successive British administrations. The cumbersome and tedious process by which such monies were authorised had led to stagnation in one of Gibraltar's main areas of employment: the construction industry.

It was against this troubled background that Señor Moran visited London in March for talks with Margaret Thatcher, and Mr Pym: talks which were seen as a preliminary to the opening of formal Anglo-Spanish negotiations on Gibraltar.[5] Señor Moran's visit was endowed with some urgency in Madrid where it was thought that, should the British Government call an early general election, there would be a hardening of attitudes on the Gibraltar question. Hence, it was felt to be imperative that the British and Spanish Governments should reach prior understanding on how the Lisbon Agreement was to be interpreted; after all, it was matters of interpretation, especially the lack of consensus on the interpretation of Clause 3, which had so far prevented any steps being taken towards the implementation of the Agreement. When the talks ended the same disagreements as had earlier existed were still unresolved and much in evidence. The Spanish position remained that Gibraltar should be restored to Spanish sovereignty through negotiation between Madrid and London whilst Britain remained adamant that the wishes of the Gibraltarians should be decisive in determining the future sovereignty of the Rock.

A futher down-turn in Anglo-Spanish relations occurred early the following month when the Spanish Government was advised of an impending five day visit to Gibraltar by the aircraft carrier HMS *Invincible* and other Royal Navy ships.[6] The Ministry of Defence described the visit as 'regular, annual and routine' with the ships *en route* to a NATO exercise, Spring Train, in the Atlantic Ocean. The Spanish Foreign Ministry greeted this news by announcing that the Government would take the 'necessary diplomatic and political measures' to ensure that Spanish interests 'will not be prejudiced' in what it regarded as its territorial waters. The British Ambassador, Sir Richard Parson, was twice summoned to the Foreign Ministry to receive protests over the visit.[7] The Spanish Government expressed 'deep concern and displeasure' at the effect that the visit of a 'great number of ships' was likely to have upon Spanish public opinion and made reference to 'events which have taken place beforehand'; obviously meaning the recent conflict between Britain and Argentina in the Falkland Islands in which HMS *Invincible* had played a prominent role. Señor Moran described the visit as 'inopportune'. However, it went ahead and the British ships sailed into Gibraltar under the watchful eyes of two Spanish frigates and a destroyer which had been despatched from Cadiz and had taken up station in Spanish territorial waters off Algeciras. In Gibraltar, the Chief Minister, Sir Joshua Hassan, said that the Spanish protests did nothing to facilitate the good neighbourliness that was a prerequisite for the implementation of the Lisbon Agreement. That good neighbourliness was not in vogue was made apparent in Madrid where Dr Manuel Fraga Iribarne, leader of the main opposition party, the Popular Alliance, stated that if he were in power he would not only immediately despatch the Spanish fleet to Algeciras Bay but also renounce the Lisbon

Agreement because it was impossible to negotiate with a British Government so insensitive to Spanish opinion. In the Senate, the upper house of the Cortes, a resolution supporting the Spanish Government's protests, and deploring a visit which was seen as worsening Anglo-Spanish relations, was unanimously carried.

The Spaniards were not alone in protesting over the arrival of the Royal Navy ships at Gibraltar for action against the ships was also taken in Gibraltar itself by the local branch of the Transport and General Workers' Union (TGWU) which was protesting against the closure of the naval dockyard. Water and fuel supplies were cut off from the British ships but the leader of the protest action by dockyard workers, Mr Joe Bossano, Branch secretary of the TGWU and leader of the Gibraltar Socialist Labour Party, agreed to co-operate with the Royal Navy in 'special cases' after the actions of his fellow trade unionists called forth a storm of opposition from many of the Rock's inhabitants. Sir Joshua Hassan condemned the trade union action as ill-timed and contended that it played into the hands of Gibraltar's enemies whilst Mr Wilfred Garcia, President of the Gibraltar Chamber of Commerce, pointed to the much needed revenue that a visit by such a large naval force would bring to the City and that the 'blacking' of the ships was self-destructive.

The 'blacking' of the ships for two of the five days that they were at Gibraltar did not have the effect of changing the British Government's decision to close the naval dockyard. The Government's closure plans, and the reasons underlying them, were the subject of a two page letter from the newly appointed Secretary of State for Defence, Mr Michael Heseltine, to the Gibraltar branch of the TGWU which emphasised the finality of the decision to close the dockyard.[8]

The dockyard's future was also the principal topic of discussions held between Mrs Thatcher and Sir Joshua Hassan on 29 April in London. Of particular concern to the Rock's Chief Minister were any British Governmental proposals that would help to secure the future of the dockyard when it was commercialised. Although the closure decision was again said to be irreversible, Sir Joshua was assured that, as a way of alleviating any economic hardship which Gibraltar might experience as a result of the change from public to private ownership, Britain would make available £28 million to facilitate the economic development of the Rock.

Regardless of the discord over their respective interpretations of the Lisbon Agreement it became clear that both the British and

Spanish Governments were agreed on one thing: that it would be inconceivable for Spain to enter the EC while a part of the territory of one member was in the possession of another member. Hence, something would have to be done about Gibraltar. Señor Moran made this evident during a visit to The Hague for talks on Spain's impending membership of the Community. Having stressed that the people of Gibraltar could retain their British citizenship under Spanish rule, Señor Moran said that his Government would shortly be presenting a new formula to Britain, a copy of which would be made available to the European Commission, which might provide a basis for a lasting solution to the problem which had bedevilled Anglo-Spanish relations for nearly two decades and which was ever-threatening to their future co-operation in the Community.[9] The British response to Señor Moran's statement suggested that a wide gulf still separated the two governments for, on 19 July, Mrs Thatcher told the House of Commons that it would be impossible for Spain to enter the EC until the frontier restrictions were lifted by the Spanish authorities. Continued British support for Spanish accession to the Community was thus implied to be conditional upon Spain making the first concession, namely, the reopening of the frontier, and it was further implied that the two governments were as far apart as ever in their respective interpretations of the Lisbon Agreement. In short, London was placing the emphasis on Madrid to take the first step in the implementation of the Lisbon Agreement and what Madrid appeared to be about to offer was an alternative set of proposals to those agreed at Lisbon.

In the meantime, the concern felt by the Gibraltar Government over the parlous nature of the Rock's economy could be gauged from the fact that Sir Joshua Hassan was back in London for further talks with Mrs Thatcher within three months of his previous visit.[10] The problem confronting the Gibraltar Government was whether or not to convert the naval dockyard into a commercial enterprise; for which Sir Joshua had secured the British promise of £28 million to facilitate the transformation earlier in the year.

Even with such financial support the time scale was short and on his second visit Sir Joshua was successful in securing British acceptance for a one-year postponement of the closure.[11] Mr Ian Stewart, Under Secretary of State for Defence Procurement, told the House of Commons on 27 July that immediately after its closure the dockyard would reopen as a commercial ship repair yard under the management of A. and P. Appledore International, a British company acting as agents for the Gibraltarian authorities; thus, the future role of the yard was determined and the Gibraltar Government bound by a decision made in London. The land and assets for the soon to be commercialised yard were to be handed over to the Gibraltarian authorities free of charge and the £28 million earlier agreed to would be utilised to meet the initial conversion costs, to provide working capital and to cover any operating losses, should these be incurred, during the first two years of the new venture.[12] Mr Stewart also promised the House that during the first three years of its operation the new yard would be provided with work commissioned by the Ministry of Defence on Royal Fleet Auxiliary vessels to the value of £14 million at current prices; additionally, work on other Ministry of Defence vessels was envisaged as generating another £500,000 to £1 million of orders for the reconstituted yard.

The postponement of the closure date for the naval dockyard by one year and the financial support that would be given to facilitate its tranformation and the early phases of its post-transference life did not fully placate the anxieties of the Gibraltar branch of the TGWU which observed that only 300 of the naval dockyard's workforce would be redeployed to the new venture and that this would leave some 900 workers unemployed. The TGWU was also doubtful as to the new yard's chances of survival in a highly competitive international market and concerned at the challenge that would be presented to it by the expanding ship repair facilities which had been developed in the nearby Spanish port of Algeciras. The fact that 900 dockyards jobs were to be left unprotected by the change in function had serious implications for the Rock's economy which was still suffering from the adverse economic consequences of the Spanish decision to partially reopen the frontier. This latter adversity was recognised by Señor Moran who contrasted the new prosperity to be found in the previously economically blighted town of La Linea with the stagnation of trade in Gibraltar; a direct result of the Spanish decision, 'when we decided to open the gate we did not think this would harm Gibraltar's economy. In fact, however, it has had important economic consequences'. However, Señor Moran made no mention of the new proposals of which he had previously spoken as a means of resolving the impasse over Gibraltar. He did admit that he had held talks with about twenty Gibraltarian 'doves' and conceded that it was disadvantageous to Spain 'to keep tightening the screws' on Gibraltar in detriment to its economy and that the current frontier restrictions, so much a cause of the Rock's financial plight, might be made more flexible if legislation was to be introduced to guarantee equal rights for Spaniards in Gibraltar.[13] Again it appeared that Clause 3 of the Lisbon Agreement was still a barrier to any progress that Britain and Spain might have been able to make on the question of Gibraltar's future status.

September witnessed the first meeting at ministerial level to discuss the Gibraltar problem since total disagreement had emerged over Spain's continuing frontier restrictions and claim to sovereignty over the Rock when Señor Moran had last met with Mrs Thatcher and Mr Pym, the then Foreign Secretary, in London in March. This time the meeting, described as 'exploratory', was between Señor Moran and the newly appointed British Foreign Secretary, Sir Geoffrey Howe, and took place in Madrid where both ministers were attending the second follow-up conference on Security and Co-operation in Europe. Although the three hours of talks between the two foreign ministers were depicted by a Spanish spokesman as having been conducted in an atmosphere of great cordiality and understanding, British sources were more restrained and stipulated that no decisions had been taken and that the only framework for resolving the Gibraltar question remained the Lisbon Agreement; which the British Government was ready to implement immediately. A further meeting between Señor Moran and Sir Geoffrey Howe at the UN General Assembly in New York on 26 September similarly failed to bring a resolution of Anglo-Spanish differences any nearer.

The autumn months of 1983 were dominated in Gibraltar by the question of the commercialisation of the naval dockyard. Increasingly, Gibraltarian politicians cast doubts on the adequacy of the British Government's aid package for the project. Similar concerns by the TGWU were turned into action when the Union refused to allow staff of Appledore Shipbuilders, the company due to take on the management of the yard when it was commercialised, into the premises. More especially, Gibraltarian opposition focused on a number of contentions. Firstly, that Gibraltar was being obliged, indeed forced, by the British Government to enter the ship repairing market at a time when virtually no ship repair yards were proving to be economically viable. Secondly, because of an over-capacity in world merchant fleets many governments were engaged in financing the scrapping of old ships and the construction of new ones, and by so doing were reducing the potential ship repair market. Thirdly, that competing repair yards in the vicinity of Gibraltar were heavily subsidised and with lower unit labour costs than the Rock could achieve.[14] Fourthly, the short transitional period specified for the transformation from a naval dockyard to a commercial ship repair yard, even allowing for the one-year postponement, meant that there was little prospect of the new enterprise becoming profitable in the short term and, hence, small scope for absorbing those made unemployed during the change. Finally, the knock-on effects of unemployment among the dockyard's workforce on the local retail sector were highlighted, especially at a time when there was scant potential for diversifying the Rock's economy because of the continuing frontier restrictions and the outflow of monies resulting from the frontier's discriminatory and partial reopening by the Spanish authorities.

In London, Government ministers became increasingly apprehensive that such opposition could endanger its plans and that, in consequence, several hundred more jobs would be lost in what was still the Rock's key industrial sector. They therefore sought to portray the package they had constructed as being fair, in that not only would the dockyard be handed to the Gibraltarian authorities free of charge, together with land on the fringe of the yard, and generous, in that additional security was being afforded by the pledge that the new ship repair yard would be guaranteed one-eighth of all work to be carried out on Royal Fleet Auxiliary craft over the next three years. It was further felt that this package reflected the continuing commitment of the British Government to Gibraltar for the package that was offered was considered to be one which would have found ready acceptance had it been extended to the naval dockyards at Devonport and Portsmouth.

Hence it was inevitable that when Sir Joshua Hassan sought a dissolution of the Gibraltar House of Assembly in December that the ensuing general election would be dominated by this single issue. For the first time since the closure of the frontier in 1969 the Rock's electorate was faced by a truly domestic issue rather than the wider question of Gibraltar's future status. The single-issue nature of the election served to ensure that it assumed the character of a referendum on the proposals agreed between the British and Gibraltarian Governments to convert the naval dockyard into a commercial ship repair yard; a fact attested to by the election being dubbed the 'dockyard election'.

The Association for the Advancement of Civil Rights – Gibraltar

Labour Party (AACR–GLP), the governing party, took this issue as its electoral theme under the slogan 'the only way ahead'. The party leader and Chief Minister, Sir Joshua Hassan, who had played the prominent role in the deliberations with the British Government defended the agreement which had been reached: 'I have obtained a very fair package and I feel a duty to see it implemented over the next four years. No one could have obtained a better deal.' Implicit in the party's electoral campaign was an appeal to the electorate to lend its support to the proven qualities of its veteran leader and long serving Chief Minister in a time of crisis.

The opposition party in the last House of Assembly, the Democratic Party of British Gibraltar (DPBG), which had held six seats, campaigned on a platform of renegotiating the agreement which had been worked out with the British Government and a demand for a further £5 million to be made available by Britain to facilitate the diversification of the local economy.

The third political party contesting the election, the Gibraltar Socialist Labour Party (GSLP), led by Mr Joe Bossano, was fighting a campaign that was made to measure for it. Mr Bossano's high prominence as a trade unionist within the TGWU, the dominant trade union on the Rock, and the active role he had played in the protest action against the closure of the naval dockyard gave the party its platform of advocating the dismissal of Appledore Shipbuilding from the Rock, the production of alternative plans for the yard's future and the use of the £28 million made available by the British Government to relaunch Gibraltar's economy on a sounder and more durable foundation.

The domestic nature of the single-issue election and the clear alternative choices offered to the electorate by the three competing political parties ensured a lively and arduous campaign and resulted in a turn-out of 75 per cent when the Rock's voters went to the polls on 26 January 1984; one of the highest turn-outs on record. Despite suffering a drop in its support from the 59.6 per cent of the vote which it had captured in the last general election of 1980 to 53.5 per cent in the current contest, Sir Joshua Hassan's AACR–GLP won the eight seats necessary to be returned to office. It was among the ranks of the Opposition parties that the greatest and most far-reaching changes occurred. The party which had held six seats on the opposition benches in the last House, the DPBG, failed to have a single candidate returned whilst the GSLP increased its representation in the House from one to seven. These changes led the leaders of both the governing party and the new opposition to claim support for their contrasting programmes with regard to the dockyard issue. The returning Chief Minister said that the election result was a vindication of his and his party's policies on the dockyard but conceded that uncertainty clouded the future; not only uncertainty over the dockyard itself but uncertainty also about the negotiations with the Spanish authorities over the reopening of the frontier. Mr Bossano described the depth of support extended to his party by the electorate as a resounding vote against the closure of the dockyard and predicted further debate and action on an issue which, despite the electoral outcome, he viewed as still being very much alive.

As teams of British and Spanish officials busied themselves over the technical details of their nations' respective differences over interpretations of the Lisbon Agreement in the search for a mutual understanding, the Rock passed from prominence in the public contacts between the two countries elected representatives during the next few months. Only occasionally was this calm disturbed as when, in July, Spain protested to Britain over the 'very marked' increase in Royal Air Force exercises by aircraft based at Gibraltar. It was alleged that RAF planes had frequently violated Spanish air space; and air space over the Rock was regarded by Madrid as its own in accordance with its claim to sovereignty. The question of air space and its possession was one of the issues which had presented an obstacle to the resolution of Anglo-Spanish differences arising from the Lisbon Agreement. A counter British protest a week later that Spanish aircraft were over-flying Gibraltar and creating a potential hazard to civilian traffic using the Rock's small airfield was not accepted by the Spanish Government. Also helping the relative tranquillity was the Spanish concentration on negotiations for membership of the EC which, delayed for far longer than had originally been envisaged, were now entering their concluding phase although difficulties relating to the free movement of labour still had to be surmounted as did the question of the length of time of the transitional period leading to it; this had been set at seven years by the Community but Spain was asking for a review after five years.[15]

Also assuming an ever-growing domestic priority was the more politically sensitive question of Spanish involvement in NATO. When the Socialist Party had come to office in Madrid as a result of the 1982 general election it had promised a referendum on Spanish membership of NATO, which Spain had entered in May 1982.

At that time Spain had, however, only committed itself to being a member of the political side of the Organisation and, despite support for full military integration among a majority of senior armed forces officers who had benefited from access to military intelligence furnished by the other member nations, the Socialist Government had stopped military integration into the command structure when it first assumed office. Whilst still not contemplating either total or partial military integration the Government was, none the less, still pledged to allow the electorate to decide whether Spain was to remain in the Organisation or to withdraw from it.

For the time being, with Spain remaining in NATO and with the prospect of membership of the EC obliging the Spanish Government shortly to contemplate the dismantlement of all the remaining frontier restrictions on the movement of people and goods between the Spanish mainland and Gibraltar, it appeared that Madrid was finding it difficult to establish any consistent momentum for its claim to sovereignty over the Rock. A chance to overcome the inertia and to reintroduce a dynamic element came in September when Britain and China reached agreement on the future of Hong Kong; with the Colony being scheduled to be returned to Chinese sovereignty in 1997. Señor Moran saw the Anglo-Chinese accord on Hong Kong as furnishing a precedent for the resolution of Spain's claim to sovereignty over Gibraltar or, more pragmatically, as a means of sensitising the views of the British to the prospect of relinquishing their possession of the Rock.[16]

Early in November came the first sign that the teams of British and Spanish officials, who had been working on the technical differences which had arisen from the differing interpretations given to the Lisbon Agreement by their respective Governments, had reached a solution when rumours began to circulate that Britain was prepared to concede the right of a limited number of Spaniards to work and buy property in Gibraltar under the transitional arrangements for Spanish entry to the EC; but that it would not permit the mass importation of Spanish labour which the Gibraltarians so deeply feared. It was further surmised that an announcement to this effect would be made later the same month in Brussels where the British and Spanish foreign ministers were to attend a special EC Council of Ministers' meeting which had been called to clarify the terms for Portuguese and Spanish entry to the Community. Those rumours proved to be well founded for, on 27 November, the two ministers concluded what was henceforth to be known as the Brussels Agreement. Unusually, this Agreement went somewhat farther than the rumours had suggested it might when, for the first time, Britain specifically consented to discuss the sovereignty of Gibraltar.

The Brussels Agreement clarified and reactivated the Lisbon Agreement which had been the subject of such differing inter- pretations and had, in consequence, complicated and delayed Anglo- Spanish negotiations on Gibraltar's future status and also delayed the full reopening of the frontier between the Rock and the Spanish mainland. The Agreement further removed a major barrier to Spanish entry to the EC, the frontier restrictions, and allowed, as earlier rumours had suggested, certain rights to be accorded to Spanish citizens in Gibraltar, such as the right to work and buy property there, one year in advance of Spain's proposed entry into the Community.[17]

The Agreement covered:

1 The way in which the Spanish and British Governments will apply by not later than 15 February 1985, the Lisbon Declaration of 10 April 1980, in all its parts. This will involve simultaneously:
(a) The provision of equality and reciprocity of rights for Spaniards in Gibraltar and Gibraltarians in Spain. This will be implemented through the mutual concession of the rights which citizens of EC countries enjoy taking into account the transitional periods and derogations agreed between Spain and Gibraltar. As concerns paid employment, and recalling the general principle of Community preference, this carries the implication that during the transitional period each side will be favourably disposed to each other's citizens when granting work permits. The necessary legislative proposals to achieve this will be introduced in Spain and Gibraltar.
(b) The establishment of the free movement of persons, vehicles and goods between Gibraltar and the neighbouring territory.
(c) The establishment of a negotiating process aimed at overcoming all the differences between (Spain and Britain) over Gibraltar and at promoting co-operation on a mutually beneficial basis on economic, cultural, touristic, aviation, military and environ- mental matters. Both sides accept that the issues of sovereignty will be discussed in that process. The British Government will fully honour the wishes of the people of Gibraltar as set out in the preamble of the 1969 constitution.

2 Insofar as the airspace in the region of Gibraltar is concerned, the Spanish Government undertakes to take the early necessary action to allow safe and effective air communications.

3 There will be meetings of working groups, which will be reviewed periodically in meetings for this purpose between the Spanish and British Foreign Ministers.

In a Note which accompanied the Agreement the Spanish Government clarified the questions which it felt that the British Government had consented to deal with regarding the Rock's sovereignty. These included 'both the theme of sovereignty of the territory referred to in the Treaty of Utrecht, as well as sovereignty of the isthmus, which was never ceded to Britain'.

In reality, neither the British nor the Spanish Government had undertaken in the Brussels Agreement to do anything that it would not have had to have done when Spain became a member of the EC. As a member Spain would have been obligated to remove the frontier restrictions it had imposed upon Gibraltar and Britain would have had to permit Spaniards the right of free movement. Even the preparedness of the British Government to discuss the question of sovereignty had only been publicly stated for the first time but had already been implicit in the Lisbon Agreement where both Governments had consented to discuss any topic raised by the other in relation to Gibraltar; clearly this could have covered the question of the Rock's future sovereignty.

Señor Moran described the Agreement as the 'biggest diplomatic success for Spain over the Rock since 1713', when Gibraltar became British under the Treaty of Utrecht, but stressed that Spain had 'the greatest respect for the feelings of the Gibraltarians themselves'. Sir Geoffrey Howe emphasised that it was now important to take advantage of the Agreement so that the economy of Gibraltar could be developed. He also pointed out that he had maintained close contact with Sir Joshua Hassan throughout the negotiations and had received the Chief Minister's support for all aspects of the Agreement. For his part, Sir Joshua described the Agreement as an 'honourable outcome' to lengthy negotiations and the first step towards fruitful co-operation between Gibraltar and the surrounding region of the Spanish mainland; the Campo.

Such positive sentiments were not forthcoming from Mr Joe Bossano, leader of the Opposition GSLP and Secretary of the TGWU branch in Gibraltar. He said that if his party came to power it would not consider itself to be bound by the Brussels Agreement and would disown it. These remarks were in keeping with his party's previous opposition to the provisions of the earlier Lisbon Agreement which the Brussels Agreement now sought to apply. More especially, Mr Bossano pledged himself to protect the jobs of some 2,000 Moroccans, constituting about one-third of all TGWU members on the Rock, who had taken over from the Spanish labour force when the frontier had been closed in 1969, and who were now threatened with unemployment if cheaper Spanish labour was allowed back into Gibraltar.

11

PARTIAL RESOLUTION

Two events guaranteed to fuel existing anxieties on the Rock, to generate heated debate and to encourage considerable speculation followed each other with great rapidity as 1985 opened. On 1 January the former naval dockyard was reborn in the guise of a commercial ship repair yard as its stay of execution expired. Two days later it was announced that Gibraltar's frontier with Spain would be fully reopened on 4–5 February, for the first time since partial restrictions had been brought into effect by the Franco Government in 1964, in accordance with the stipulations contained in the Brussels Agreement.[1]

To resolve the final details of the unrestricted reopening of the frontier between the Rock and the surrounding Spanish mainland talks were held, on 10 January, between Spanish and British representatives, the latter being assisted by Gibraltarian officials, at the town hall in La Linea. The previous day a meeting chaired by Señor Jose Rodriguez de la Borbolla, Chief Minister of the Andalucian Regional Government, had been attended by the seven mayors of the Campo region at which they had been assured that the local economies of the area would be protected against any adverse effects which the free passage of people and goods across the frontier might produce.[2] Concern at the economic consequences of a fully open frontier was thus shown to be something which was not peculiar to the inhabitants of the Rock and although the Campo towns had benefited disproportionately from the partial reopening of the frontier in 1982 there were many who questioned whether such advantage was anything more than a short-term phenomenon which would disappear with the withdrawal of all frontier restrictions. This concern was voiced by the Mayor of La Linea, Señor Antonio Diaz, who called for an immediate cash injection by the Madrid

Government equivalent to £2.5 million.

The day following the meeting in La Linea an equivalent gathering was held in Gibraltar where the British delegation, led by Mr John Broadley, Gibraltar's Deputy Governor, and the Spanish team, under the leadership of Señor Francisco Mayans, special adviser to the Spanish Foreign Minister, Señor Moran, discussed further the practicalities of unhindered communications, customs and passport controls, car parks, currency exchanges and bus and taxi services. The two rounds of talks were described by Señor Mayans as having ended satisfactorily although certain points of detail had been referred back to the two national governments.[3]

In Clause 1(a) of the Brussels Agreement it had been stipulated that legislation would be introduced in both Spain and Gibraltar to ensure 'the provision of equality and reciprocity of rights for Spaniards in Gibraltar and Gibraltarians in Spain'. It was to the enactment of that legislation that the Gibraltar House of Assembly turned its attention on 16 January. A Bill which gave rights of entry and residence, land purchase, the establishment of businesses, family allowances and emergency medical treatment to Spaniards in Gibraltar was presented to the House by Gibraltar's Chief Minister, Sir Joshua Hassan. He contended that since the issue of Gibraltar's sovereignty was 'a totally unrelated question' to that by which the House was now confronted, Gibraltarians should take advantage of the reopening of the frontier with Spain one year before Spain's anticipated date of entry to the European Community.[4] He stressed that the free movement of labour was a question for the future as Community legislation imposed a 7-year transitional period between membership and the ending of all restrictions on the mobility of workers. The Deputy Chief Minister, Mr Adolfo Canepa, argued that a partially open frontier had been shown not to be tenable and that if the House did other than act in accordance with the requirements of the Brussels Agreement then Gibraltar ran the risk that Britain and Spain would decide its future over its Government's head. Speaking against the Bill, Mr Joe Bossano leader of the Gibraltar Socialist Labour Party, which had denounced the Brussels Agreement, described the Bill as 'the most shameful piece of legislation ever presented to Gibraltar's House of Assembly' and challenged the Government to hold an election on the whole issue of the Agreement.

After heated debate and acrimonious exchanges between Government and Opposition, the Bill was approved by 8 votes to 7.[5]

When the Bill reached its committee stage and third reading the Opposition boycotted the proceedings. On 31 January, the Spanish Government formally approved the removal of all restrictions on Gibraltar and agreed to allow the free passage of persons, vehicles and merchandise between Spain and Gibraltar together with the resumption of direct passage by boats; as it was required to do under the Brussels Agreement.

In Gibraltar continued opposition to the Brussels Agreement was not only manifested in the House of Assembly but also on the streets where a peaceful demonstration by some 1,300 Gibraltarians signified their objection to it. Mr Joe Bossano presented a letter carrying 9,426 signatures to the Governor of Gibraltar, Admiral Sir David Williams, and claimed that this reflected the substantial opposition felt by the people of the Rock to the Agreement.[6]

Despite the reservations and protests on the Rock, at one minute after midnight on 5 February, the Spanish Governor of Cadiz, Señor Mariano Baquedano, removed the padlocks and unbolted the gates on the Spanish side of the frontier with Gibraltar and by so doing ended the long period of isolation. Gibraltar had withstood another siege in its illustrious history but what could be seen as a famous victory for the resolve and stoicism of its citizens was not greeted by them with joy but with a nervous concern. It was perhaps to allay these concerns that the Governor of Gibraltar called for patience in judging the changes occasioned by the Brussels Agreement; Admiral Sir David Williams said that in his judgement 'bearing in mind the solemn undertaking of the British Government on sovereignty, the future prosperity of Gibraltar will be enhanced by what happens tonight'.

In accordance with the Brussels Agreement, the British Foreign Secretary, Sir Geoffrey Howe, and the Spanish Foreign Minister, Señor Moran, met in Geneva on the same day as the frontier was reopened. Also present at that meeting was Sir Joshua Hassan who had earlier travelled to London for talks with Sir Geoffrey, in advance of the Geneva meeting, at which he had made clear to the Foreign Secretary that he could not support any process which might lead to the transfer of Gibraltar's sovereignty to Spain during the next two generations.[7]

Although the question of sovereignty was raised by Señor Moran it was not discussed in depth and again the Spanish Foreign Minister hinted that Spain would later present formal proposals on that topic. Instead, the two Foreign Ministers satisfied themselves with a consideration of less contentious matters and agreed detailed procedures for discussing issues relating to Gibraltar which entailed regular annual meetings of foreign ministers to deliberate upon questions of mutual interest, including that of Gibraltar's future status. They also signed an agreement on economic and cultural co-operation and established working parties of British and Spanish officials which were to immediately start work to promote bilateral understanding on other matters felt to be important and divisive.

On the same day as the Geneva meeting, the British Prime Minister, Mrs Thatcher, told the House of Commons that 'the Government cherished the freedom of the people of Gibraltar to decide their own future . . . [and] has given assurances . . . that the Government will never enter into arrangements in which the people of Gibraltar will be brought under the sovereignty of another state against their freely and democratically expressed wishes'. However, as Señor Moran observed, this statement did not preclude the possibility of a negotiated transference of the Rock's sovereignty.[8]

The ending of the frontier restrictions heralded an immediate economic boom in Gibraltar. Within the first week of the reopening an estimated 45,000 people visited the Rock and trade and commerce flourished. The President of the Gibraltar Chamber of Commerce, Mr Jimmy Risso, expressed pleasure that he had publicly given his support to Sir Joshua in the days prior to the Geneva talks and Mr Haresh Budhrani, Secretary of the Indian Merchants' Association, said that although 'it's early days, any businessman who says he can't earn a decent living in Gibraltar today is a born failure'. The rapid increase in banking and financial services and a decline in unemployment confirmed the remarkable turn-around in the Rock's economy. Doubts, however, were still to be found among those who realised that the question of sovereignty remained, as yet, unresolved. The Gibraltar Socialist Labour Party continued to insist that the reopening of the frontier would prove to be disastrous for the economy in the long term as the current euphoria overlooked both its lack of diversity and the fact that to date no work permits had been issued to Spaniards seeking employment on the Rock.[9]

One of the central elements in the Rock's future prosperity was its airport which was alone capable of providing the ease of access so imperative to the development of a tourist industry; an industry regarded by many as a necessary prerequisite to financial independence and long-term economic vitality. Under the Brussels Agreement, Spain had promised to take the 'early actions necessary to allow safe and effective air communications'. In the Note which accompanied the Agreement the Spanish Government had also clarified the questions that it felt Britain had consented to discuss in connection with Gibraltar's sovereignty; among these had been 'the sovereignty of the isthmus, which was never ceded to Britain'; Gibraltar airport straddled the isthmus between the Rock and the Spanish mainland. To address this issue teams of British and Spanish officials met on 4 March, with Britain wanting Spain to lift its restrictions on flights over the Bay of Algeciras so that Gibraltar airport could be more fully utilised. This request was granted under new regulations published in Madrid on 1 April.[10]

It was to Gibraltar airport that Sir Geoffrey Howe flew on 7 June, direct from a meeting of NATO foreign ministers in Lisbon where he had held discussions with Señor Moran on the progress made over the Rock since the signing of the Brussels Agreement.[11] On 10 June, Sir Geoffrey told the House of Commons that he had reiterated the Government's pledge to respect the freely and democratically expressed wishes of the people of Gibraltar and that he had been 'heartened to hear about the Gibraltarian Government's plans to diversify and strengthen their economy and to develop practical co-operation with neighbouring Spain in a way which would be of benefit to both'. Closer links between Gibraltar and the towns of the Campo region had been initiated in March when Señor Rafael Palomino, President of the recently created 'Mancommunidad' of the municipalities of the Campo, had visited Gibraltar and had been further advanced when Sir Joshua Hassan had visited Algeciras in April; the first visit by a Gibraltarian Chief Minister to Spain since the imposition of frontier restrictions in 1964. Whilst in Gibraltar, however, Sir Geoffrey had told the Gibraltarians that the option of independence was not one open to them because the British title to sovereignty of the Rock was founded in the Treaty of Utrecht and that Treaty provided for the return of sovereignty to Spain if British sovereignty ended as a result of the wishes of the people of Gibraltar.

It was against this background of a Spain eagerly awaiting entry to the EC that Sir Geoffrey Howe visited Madrid on 4-5 December for the second of the periodic discussions between the foreign ministers of Britain and Spain stipulated in the Brussels Agreement.[12] In a joint communiqué issued after the talks, Sir Geoffrey and the recently appointed Spanish Foreign Minister, Señor Fernandez Ordonez, stated that their talks on Gibraltar had encompassed a full discussion of the issue of Gibraltar's sovereignty, that they had reviewed the Spanish Government's proposals of February 1985, and had agreed that the study of the sovereignty issue should continue through diplomatic channels. Significantly, the exact nature of the Spanish proposals on Gibraltar's future status was not made public and, in consequence, this led to considerable speculation in the Spanish press where it was rumoured that the proposals involved a 'lease back' arrangement along the lines of that agreed between Britain and China over Hong Kong or some form of Anglo-Spanish condominium for the Rock.[13] Also agreed by the two foreign ministers was 'the common objective of developing the civilian use of Gibraltar's airport on a mutually beneficial basis' while Sir Geoffrey confirmed that Spanish pensioners with claims arising from previous employment in Gibraltar would be paid at the same rate as Gibraltarians from 1 January 1986; the date of Spanish entry into the EC.

Hopes that 1986 would be the year in which real progress would be made towards a final and lasting solution to the Rock's future status, acceptable to all involved, were voiced by Señor Ordonez in March. Addressing a gathering of foreign correspondents in Lisbon, after a meeting with Portuguese political leaders on the means by which the two countries could work together in the EC and NATO, he said that the problem posed by Gibraltar to Anglo-Spanish relations would be easier to solve 'now that Spain and Great Britain are friends within the EC and NATO and will be sitting at the same table discussing together'. He did, however, point out that he viewed Gibraltar as 'the only colony in Europe'; a situation which he described as 'anachronistic and morally intolerable'.[14]

Such sentiments were to be proved to be both a little optimistic and a little premature, for the British Government alleged that an incursion had been made into Gibraltar's territorial waters by the Spanish aircraft carrier, and flagship of the Spanish Navy, *Dedalo*, on the night of 20–21 March and that two helicopters had been launched from the ship's deck into the Rock's airspace.[15] In response, the Spanish Foreign Ministry stated that the Treaty of Utrecht recognised as Gibraltar's territorial waters only the immediate area of the port and that the surrounding waters remained Spanish.

This incident was not permitted to lessen the enthusiasm surrounding the visit of King Juan Carlos and Queen Sofia to Britain on 22–5 April. At a state banquet the King said of Gibraltar that he was certain that 'our two governments will find, by means of a negotiating process already begun, appropriate formulae for reaching a solution satisfactory to all.'[16] He was to repeat this theme when he addressed both Houses of Parliament the following day:

> The recently resumed dialogue over Gibraltar is a step forward, but there remains a long way to go. I trust our respective governments may be capable of . . . [transforming] any shadow into an element of harmony for the greater co-operation between our countries and the general well-being of the interested parties, as well as the future of Europe.[17]

Nor did King Juan Carlos change the tenor of his remarks when he addressed the 41st Session of the United Nations General Assembly in New York later in the year; but divisions remained.[18] Foremost among these was the question of Gibraltar's airport which came to dominate Anglo-Spanish discussions throughout the year.

In early February, British and Spanish officials gathered in Madrid to consider wider civilian use of Gibraltar airport and the prospects for a direct air link between the Rock and the Spanish capital.[19] Problems had arisen over two questions: namely, Spain's refusal to grant approval to flights through Spanish airspace by British military aircraft using the airport, and the status to be accorded Spanish citizens using the airport before crossing into Spain. A further round of talks on these subjects, set for 7–8 June, was postponed at the request of the Spanish Government but the threads of the debate were again picked up on 10 September in Madrid where it emerged that the talks had encountered technical and political hurdles. The technical hurdle primarily related to the current and future envisaged processes of airport utilisation but the political hurdle arose from Spanish demands that two terminals should be established at Gibraltar airport: one specifically for flights to and from Spain, and under Spanish control, which would allow passengers to avoid the Rock's customs and passport controls. To this, the Gibraltarian Government was opposed and, on 17 December, the Gibraltar House of Assembly unanimously supported a resolution which called for the airport to remain exclusively under the control of the British and Gibraltarian authorities.

Hopes that the remaining difficulties standing in the way of an Anglo-Spanish resolution of the Gibraltar question were expressed by the Spanish Prime Minister, Señor Felipe Gonzalez, when, having been returned to office, he presented his new Government's programme to the Cortes on 23 July. He contended that Spain had a firmer basis than in the past for pursuing its foreign policy objectives now that it had joined the EC and clarified its membership of NATO through the referendum which had been held in March. With these two major elements of foreign policy having been resolved, Señor Gonzalez said, 'We hope that in this legislature we can advance towards, or reach, a definitive solution over decolonisation, over the mechanism for decolonising Gibraltar'.

What had the outward appearance of a growing affinity between London and Madrid when viewed from Gibraltar was not greeted there with warmth. The question of the utilisation of Gibraltar airport had already served to raise doubts on the Rock that the British Government seemed prepared to offer concessions to its Spanish counterpart with nothing being granted in return. When, at the end of July, the British Government unexpectedly withdrew the ceremonial military guard posted at the frontier separating the Rock from the Spanish mainland, without a reciprocal gesture by the Spanish authorities, this was taken as a further sign that the British were no longer thinking of maintaining a long-term presence on the Rock; with the benefit of hindsight, the commercialisation of the naval dockyard came to be viewed as having been the first step along the road to a British withdrawal from the Rock.[20]

Despite these alarums there were also signs that Gibraltar had benefited from the Brussels Agreement. The trade figures for 1985 showed that, since the frontier with Spain had been reopened in February of that year, the Rock's exports had grown by 72 per cent, £14.5 million more than in the previous year, and that imports had risen, although less sharply, by 58 per cent.[21] Tourism had generated £20 million for Gibraltar as some two million visitors came to the Rock in the twelve-month period since the normalisation of the frontier compared to an average annual figure of only £150,000 in each of the three previous years. In short, the Gibraltarian economy was undergoing something of a boom even though doubts remained as to its long-term strength. These doubts were given substance by the revelation of the Gibraltar Government's economic adviser who indicated that Spain had made a greater profit out of the reopening of the frontier than had the Rock itself; exports from Spain to Gibraltar doubled in the first quarter of 1985.

Also of concern was the financial burden incurred by the payment of pensions to elderly Spanish citizens who had spent much of their working lives in Gibraltar prior to the closure of the frontier in 1969. In the first two months of 1986, hundreds of Spaniards had queued in Gibraltar to collect pension money owed to them since Spain's accession to the EC. The payment of those pensions was estimated to inflict a cost of £7 million on Gibraltar's economy, annually, for the next fifteen years.

This amount would have to be found in addition to an annual expenditure commitment which was estimated at £15 million for 1985. For the present much of this extra cost was being met by the British Government but there were doubts raised as to how long this arrangement would continue before Gibraltar was left to bear the full weight of the payment of pensions to Spanish citizens.[22]

If the pace and thrust of Anglo-Spanish negotiations on Gibraltar's future, and what was seen as a too lenient and tolerant a stance being assumed by the British Government, was occasioning consternation on the Rock, the opposite was seen as a cause of concern in Madrid where Señor Jesus Ezquerra, Director-General of the European Department of the Spanish Foreign Ministry, stated, on 17 December, that 'We note a clear intransigence by Britain to negotiate with Spain.' Similar disappointment was voiced by Señor Ordanez who professed himself less than happy with the progress which had been made on the question of Gibraltar's future since the frontier restrictions were lifted in February 1985. Attention, therefore, came to be sharply focused on the meeting of the British and Spanish foreign ministers which was to take place in London on 13–14 January 1987, to see what was the current state of play between Britain and Spain in the contest over the Rock.

Yet even before these talks had commenced, relations between London and Madrid had been exposed to further strain. On 6 January Admiral Cesar Pellini, the NATO Commander Allied Forces Southern Europe, had visited Gibraltar in what the Spanish Government took to be an affront to its claim to sovereignty of the Rock. On 8 January, a Spanish radio interview with Mr John Grant, an official at the British Foreign and Commonwealth Office, led to more animosity when Mr Grant stated that the British Government was not prepared to make any concessions on Gibraltar's sovereignty nor to discuss any proposals which might lead to changes in its declared position on that question.[23]

Nevertheless, when the two foreign ministers met in London their discussions focused on progress in co-operation between Spain and Gibraltar, the civil use of Gibraltar airport and the question of the Rock's sovereignty. However, despite consideration of those discreet issues, which were clearly part and parcel of the Gibraltar question, a breakthrough was not achieved.[24] After the talks Señor Ordonez told a news conference that there was an 'abnormal situation' in being when one NATO member country had a colony in another country of the Atlantic Alliance and when one member country of the European Community maintained a colony in another Community member country. He contended that 'the shadow of the Rock is projected over our relations' and that the dispute over Gibraltar's future was not only a threat to Anglo-Spanish relations but also to the two nations' co-operation in both the EC and NATO; a co-operation which he described as having been 'already damaged'.[25] Perhaps in an effort to bring the pressure of NATO to bear upon the British Government, Señor Ordonez said that the Organisation's military communications facilities in Gibraltar would have to be removed because the Rock was a part of Spain and because Spanish membership of the Organisation did not involve participation in NATO's integrated military command structure.

The British response to the failure of the two foreign ministers to resolve the differences which continued to separate their respective governments in relation to Gibraltar came on 19 January when, Mr Timothy Eggar, Parliamentary Under-Secretary of State for Foreign and Commonwealth Affairs, noted, in a written parliamentary reply, that contacts had developed satisfactorily in a number of fields and that Sir Geoffrey Howe and Señor Ordonez had agreed to the creation of economic and cultural working groups but that the commitment of the British Government to honour the wishes of the people of Gibraltar, as detailed in the preamble to the Rock's 1969 Constitution, had been reaffirmed and 'the importance of managing any differences between Britain and Spain in a spirit consistent with their links of traditional friendship and their common membership of the European Community and NATO' had been emphasised.[26]

Whilst Britain and Spain deliberated Gibraltar's future the Gibraltarians had been actively engaged in securing their community's self-sufficiency. Even the initial drain upon the local economy occasioned by the partial reopening of the frontier had been surmounted and converted into the sustained economic boom which had been first evidenced in 1985 in the wake of the Brussels Agreement. Reservations were still voiced as to whether or not the current signs of prosperity were temporary or an indication of a longer-term financial viability; given the Rock's dependence upon the Spanish authorities keeping open the single access point from the

Spanish mainland and the question mark which hung over Gibraltar Ship Repair when its three year cushion against the vicissitudes of the open international market place was removed at the start of 1988.

One factor was common to both the opportunities and the threats by which the Gibraltarians were confronted in their battle for economic self-sufficiency: the Rock's geographical position. The threat was posed by the uncertainty over the future actions which the Spanish Government might take in pursuit of its claim to sovereignty and the concessions which it might wring from its British counterpart. Whilst few on the Rock envisaged a return to the black days of enforced isolation everyone knew that access over the narrow isthmus to La Linea was crucial to their continued economic well-being. The isthmus was their achilles heel; not only did it carry the road to Spain but it was also the site of the airport.

Although the magnitude of these threats and their resolution would largely be dependent upon the outcome of Anglo-Spanish negotiations which the Government of Gibraltar would continue to seek to unfluence, the opportunities were less prone to outside determination and the manner in which they could be exploited would primarily depend upon the initiative and enterprise of the Gibraltarians themselves. The Rock's geographical position of close proximity to the several hundred thousand British expatriates living on the Spanish Costa del Sol and its ready communications to the oil-rich Middle East made it attractive to companies anxious to exploit openings in both of those markets. Taking this opportunity, Gibraltar was rapidly developing into one of the world's larger fully fledged offshore financial centres.

Banks, multinational companies and building societies vied with each other to acquire premises on the Rock's limited land area as they sought to take advantage not only of its favoured location but also of its generous taxation regime.[27] Under Gibraltar's offshore legislation, companies registered on the Rock, but not trading within its territorial limits, did not pay tax on profits except for an annual fee of £225. Also, these so-termed 'exempt' companies could, unlike companies in other offshore locations, be managed and controlled from within Gibraltar itself. It was, therefore, hardly surprising that Gibraltar's Financial and Development Secretary, Mr Brian Traynor, was led to comment that 'financial services are offering us the most promising avenue to self-sufficiency'.

Paralleling Gibraltar's growth as a financial centre was the growth in its tourist industry. The number of direct flights between the

Rock and London doubled in the period 1984–7 but concern was expressed that many of the visitors to Gibraltar were either in transit or merely day trippers from the better established Spanish holiday resorts of the nearby Costa del Sol and that the larger spending power that came with long-stay tourists and businessmen was not being fully achieved. Nevertheless, the President of Gibraltar's Chamber of Commerce, Mr Sol Seruya, pointed to the benefits accruing from the Rock's emergence as a commercial centre when he noted the increase in import duty revenue which had accrued to the city's exchequer over the last three years; £4.4 million in 1984 rising to £7.7 million in 1985 and to nearly £10 million in 1986.

The Rock's Chief Minister expressed himself to be satisfied that 'parts of the economy are doing well. I think we can look forward to the future with confidence'. Indications of that confidence were to be evidenced in the planned £50 million investment by the international construction group, Taylor Woodrow, which was the prime mover in the development of Queensway Quay, on the Rock's western side, into a hotel, marina and apartment complex by 1991. The growing co-operation between the Rock and the towns of the adjoining Campo region of Spain was attested to by the discussions which were taking place between Gibraltarian and Spanish officials to develop a common marketing strategy for the entire area.

Geographical location was also of importance to the three dry docks, two repair yards and their associated workshops which collectively had become Gibraltar Shiprepair on 31 December 1984. With 6,000 vessels a month passing through the Straits linking the Atlantic Ocean and the Mediterranean Sea the yard was well placed to capitalise upon one of the world's busiest seaways. Yet the moment of truth for the ship repair yard lay ahead, at the end of 1987, when the work on Royal Fleet Auxiliary craft, earlier guaranteed by the British Government at the time of the commercialisation of the former naval doakyard, expired. Disruption in the ship repair yard over the style of management which Appledore's managing director had introduced had led to strike action being taken by the workforce in 1986 and his replacement by an experienced Scandinavian ship builder.[28] Again on the positive side, orders had been secured from West German and Scandinavian companies and a feasibility study commissioned by the Gibraltar Government and carried out by the international accountants, Price Waterhouse, had indicated that, subject to some necessary and specific expenditure, the yard's future appeared to be one of economic viability. Such optimistic projections were confirmed mid-way through 1987 when it was revealed that the first six months of that year had been the most profitable since the yard had been commercialised. However, strike action by the workforce in August led to the further revelation that the yard was in imminent danger of closure. Selective stikes were instigated by industrial workers in search of parity with their equivalents in Britain because they feared that overtime would become scarce once the yard had to compete in the open market from the end of the year. This resulted in a demand for a 40 per cent pay rise which stood in marked contrast to the management offer of 6 per cent and the workers rejected a call from their union to return to work while negotiations took place to see if some way could be found to bring the two sides closer together. To add to the uncertainty, A. and P. Appledore, the yard's management, had been given notice of default by the board of Gibrepair because it had failed to provide a satisfactory computing system and adequate information on accounts. It was the board's hope that A. and P. would consent to renegotiate its 10-year management agreement so as to permit the board, which had the Gibraltar Government as its only shareholder, more say in the running of the yard.[29]

Gibraltar's capacity to influence not only the course of Anglo-Spanish relations but also wider European developments was high-lighted in June 1987 when European aviation policy was thrown into disarray after Spain raised last minute objections to an agreement which would have introduced air transport deregulation and cheaper air fares throughout the Community.[30] Spanish objections arose because, under the terms of the agreement which was being sought, Gibraltar airport would have been treated as a British regional airport under exclusive British control and would, accordingly, have benefited from the deregulation measures for regional airports; this clearly adversely affected Spain's claim to sovereignty over Gibraltar.

Negotiations between the British Secretary of State for Transport, Mr Paul Channon, and the Spanish Minister for Transport and Communications, Señor Abel Ramon Cabballero Alvarez, failed to save the proposed agreement by the deadline of midnight on 30 June; when, with new EC decision-making rules under the Single European Act coming into force at that time, the consumer oriented air fares policy would have to be subjected to a long process of scrutiny by the European Parliament before it could be finally agreed. Commenting on events in Luxemburg, the Spanish Prime Minister said that he did 'not believe this needs to happen permanently' and that he had recommended that it would be advisable in the future if solutions were sought to problems relevant to Britain and Spain at the bilateral level and before they reached the Community level. However, he insisted that on this occasion Spain's vital interest in the question of Gibraltar's sovereignty had to be appreciated but he did not rule out the possibility that Spain would again invoke a 'vital national interest' over the Rock in order not to be out-voted by majority decisions now that the Single European Act had come into force. In Britain the events which had taken place in Luxemburg and the use of the Spanish veto led to concerns that having won on the question of air transport deregulation Spain might have created a precedent and might be prepared to apply new economic pressure on Gibraltar; whilst London was seeking acceptance for its view that Gibraltar could not be excluded from the benefits of EC membership.

The European spotlight remained on Gibraltar when, on 16 September, a delegation from the Rock, led by Sir Joshua Hassan and Mr Joe Bossano, visited the European Parliament in Strasburg to address meetings of any interested Euro-MPs with the aim of explaining Gibraltar's opposition to the Spanish claim to sovereignty over the city. This visit took place in the face of the contention by Madrid that any settlement of the Gibraltar question must emanate from direct negotiations between Spain and Britain.[31]

It was against this background of an increasingly vociferous and pro-active Rock that three rounds of Anglo-Spanish talks were held in October to try to resolve the Gibraltar airport issue. The strength of feeling and the depth of opposition in Gibraltar to any concessions being made to Spain during Anglo-Spanish talks on the airport were made evident on 10 November when a large public demonstration was organised on the Rock during a two-day visit by Mr David Radford, a Foreign and Commonwealth Office official, who was holding technical talks with Spain on cross-frontier co-operation. News of Mr Radford's visit had again caused rumours to circulate among the Gibraltarian community that British sovereignty was imminently threatened. The fact that Mr Radford had met with local politicians, businessmen and trade unionists to explore a proposal to allow passengers from Spanish departure points to fly to Gibraltar and walk into the neighbouring Spanish town of La Linea without passing through British immigration and custom controls did nothing to dampen the conjecture. The demonstration had been organised by Action for Gibraltar, a group formed earlier in November by six civil servants and the Gibraltar Trades Council, and was described as the biggest in Gibraltar for twenty years. A demonstration headed by the Chief Minister and the Leader of the Opposition and all of the representatives in the House of Assembly was followed by some 12,500 local inhabitants and a 'no concessions to Spain on the airport' petition bearing 15,500 signatures was handed to Mr Radford to be passed to the British Foreign Secretary.

Assured of the support of the people of the Rock, the Gibraltar House of Assembly unanimously passed a resolution calling upon the British Government not to make any concessions to Spain on the airport issue and, at the same time, the Gibraltar Trades Council, representing three-quarters of the city's workforce, voted to take direct action to prevent any future Anglo-Spanish agreement on the airport being implemented if it was not in accord with Gibraltarian opinion.[32] Señor Gonzalez responded with an equal display of resolve when, on 21 November, he again emphasised that Gibraltar airport was built on land not ceded by Spain under the Treaty of Utrecht and that if his Government accepted any EC directive on the airport this would constitute the first legal recognition by Spain of a change in the status of the territory.

The airport issue had now become the one which, at least for the present, dominated all others, and had assumed a particularly pressing significance as, under the Brussels Agreement, the British and Spanish foreign ministers were shortly scheduled to meet for their annual round of talks. Therefore, Sir Joshua Hassan flew to London for consultations with Sir Geoffrey Howe and two days later Sir Geoffrey and Señor Ordonez met in Madrid in an attempt to settle the question of the Rock's airport. Speculation surrounded the Madrid meeting and it was rumoured that the British Foreign Secretary was about to offer to Spain what the Gibraltarians were bound to regard as a concession, namely, the joint use of Gibraltar's airport free from immigration and customs controls. Were such a concession to be made, it was then anticipated that Spain would withdraw its objections to the British-promoted EC plan for the deregulation of European air transport. By linking the European air agreement, which had been proposed by the EC transport ministers at their June meeting, with the unrelated issue of Gibraltar airport, Spain had obliged Britain to seek the middle ground between the Spanish demands for joint control of the

Rock's airport and the Gibraltarians' opposition to anything that remotely resembled a concession to Madrid on the wider question of sovereignty. In consequence, the British had been obliged to offer a scheme along the lines of that earlier discussed by Mr Radford during his investigative discussions in Gibraltar. Sir Geoffrey Howe had since argued that such a modus vivendi, which met neither the full demands of the Spanish nor of the Gibraltarians, 'need involve no infringement of sovereignty whatsoever'; a view certainly not supported on the Rock where such an arrangement was perceived to be another step along the road towards the abandonment of British sovereignty.

The Madrid talks failed to resolve the question and Señor Ordonez contended that technical problems of civil aviation together with legal and political difficulties still stood in the way of a solution. The urgency with which the airport question was viewed by both the British and Spanish Governments led to a further meeting of the two foreign ministers, in London on 2 December, at which agreement was finally reached on co-operation over Gibraltar airport as well as upon the resumption of the ferry service between Algeciras and Gibraltar and also upon the acceleration of the traffic flow across the frontier between the Rock and the Spanish mainland.[33]

At the end of the London meeting a joint declaration was issued by the British and Spanish Governments which stated that:

1 The aeronautical authorities of the two countries would hold regular discussions on the civil use of Gibraltar airport, including the establishment of new services to third countries.
2 Spain would allow Spanish airlines to operate between Spanish airports and Gibraltar airport under EC regulations.
3 Spain would build a new terminal at La Linea, adjacent to the northern side of the existing frontier fence, and passengers using it would have direct access to Gibraltar through a gate in the south side of the terminal.
4 The Spanish terminal would be used by passengers going to any point north of the frontier fence and by those proceeding from this area.
5 The UK terminal would be used by all other passengers.
6 Where appropriate, passengers would be subject to customs and immigration controls in the respective terminals.
7 An Anglo-Spanish committee would co-ordinate both terminals' civil air transport activities and their relations with the airport's

other services.

8 There would be close co-operation on airport security.
9 Continued discussions would take place between the UK and Spain on further strengthening air safety and on traffic control arrangements in the area.
10 All arrangements would come into operation on condition that the Gibraltar House of Assembly passed the necessary legislation to give effect to the provision regarding customs and immigration controls (i.e. to allow passengers who would use the Spanish terminal to be exempted from Gibraltar's customs and immigration controls) and not later than one year after the UK notified Spain that this had been done.

On sovereignty, the declaration added that 'the present arrangements and any activity or measure undertaken in applying them or as a consequence of them are understood to be without prejudice to the respective legal positions of Spain and the United Kingdom with regard to the dispute over sovereignty over the territory in which the airport is situated'.[34]

Sir Geoffrey Howe described the agreement which had been reached as providing for 'practical and co-operative arrangements' and contended that 'these arrangements do not impair British sovereignty over Gibraltar which is fully preserved'. Both sides had made concessions. The British had agreed to give Spain joint use of Gibraltar airport for civilian, but not for military, purposes and Spain had abondoned an earlier demand for joint administrative control of the airport in return for joint usage.[35] Furthermore, the Anglo-Spanish agreement had seemingly removed the last obstacle to an EC air transport liberalisation directive.

The agreement, however, had been produced at a price and the person who was to pay that price was Gibraltar's Chief Minister. Just over a week after the joint declaration was announced, Sir Joshua Hassan tendered his resignation.[36] Although he asserted that his departure from Gibraltar's political life was for personal reasons, and not because of the Anglo-Spanish agreement on the Rock's airport, there were many who claimed to detect a causal link between the two events. On the same day, Mr Adolfo Canepa, who had hitherto been Deputy Chief Minister, was sworn in as Sir Joshua's successor.

The airport issue was not, however, about to pass into history for, on 17 December, the Gibraltar House of Assembly effectively rejected the Anglo-Spanish agreement on its joint usage when it unanimously voted to challenge the legality of the Rock's exclusion from the EC arrangements on air transport deregulation and also by approving a resolution reiterating Gibraltar's right to be included in the EC measures 'as a regional British airport without preconditions'.

12

AN EYE TO THE FUTURE

The Gibraltar General Election of 24 March 1988 marked a watershed in the politics of the Rock. Victory for the Gibraltar Socialist Labour Party (GSLP) ended 16 years during which the Association for the Advancement of Civil Rights – Gibraltar Labour Party (AACR–GLP) had guided the Rock's affairs. When the votes were counted the GSLP emerged as the outright winner with 58 per cent of the popular vote and 8 of the 15 seats in the House of Assembly; one more than its chief rival the AACR–GLP which saw its percentage of the poll fall to an unprecedented low of 29 per cent. The newly created Independent Democratic Party (IDP), fighting its first electoral contest, trailed home a distant third; its 13 per cent of the vote proving to be too small to secure for it any representation in the new House.

A variety of factors may be said to have produced this radical departure from the customary pattern of Gibraltarian politics. Firstly, the election was the first to be held in Gibraltar since the signing of the 1984 Brussels Agreement between Britain and Spain which had ended all of the Spanish restrictions on Gibraltar's frontier with the mainland and which had made possible the negotiations on the opening of the Rock's airport to Spanish airlines and which had, in turn, led to the agreement of 2 December 1987. To this extent, the election afforded Gibraltarians their first opportunity to pass judgement on those two far-reaching and significant accords. Secondly, the resignation of Sir Joshua Hassan, as Chief Minister and Leader of the AACR–GLP, had left his party without its long-serving and widely experienced leader and entrusted its future to the less well known and relatively untried and untested Mr Adolfo Canepa. Thirdly, there was the split which had arisen in the ranks of the AACR–GLP in the wake of Sir Joshua's resignation with the formation of the IDP by Mr Joe Pietaluga; a senior civil servant who had acted as Sir Joshua's chief adviser. Fourthly, the growing prominence of the GSLP as a clearly discernible centre of opposition, to both the 1984 and 1987 Anglo-Spanish accords and the vigorous campaign which it had waged against the privatisation of the Rock's former naval dockyard, offered the electorate a clearer choice than had often been the case in previous elections to the House of Assembly. Fifthly, credence must be given to the demand for political change and, more especially, for a change in the style of political leadership which emanated from the electorate. The reopening of the frontier, the departure of Sir Joshua Hassan from political life, the growing accord in Anglo-Spanish relations and the emergence of the GSLP as an ever-growing and vibrant political alternative to the AACR–GLP under the dynamic leadership of Mr Joe Bossano, all heralded what looked like the beginning of a new era for the Rock. In consequence, the possibility cannot be discounted that the electorate sought a new political force to safeguard the community's interests in a fast-changing environment.

It should not, however, be thought that the victory of the GSLP was attributable simply to circumstance or to the actions or failures to act of its political rivals for the party also successfully capitalised upon its own strengths and its radicalism could be said to have captured the imagination of the Gibraltarians themselves and reflected most accurately the opposition that many of them felt to both the Brussels Agreement and the form taken by the resolution of the question of the utilisation of the airport. At the same time it must also be conceded that the domestic policy proposals of the GSLP accorded with the electorate's view of the need for a growth and expansion in the Rock's economy after the years of stagnation and constraint occasioned by the Spanish imposition of frontier restrictions. The GSLP's proposals for increased public expenditure which encompassed the creation of a Gibraltar national bank, a public take-over of the loss-making ship repair yard, plans for land reclamation, the intention to construct 500 units of low-cost public housing over the next four years and the development of the financial service sector of the economy to make Gibraltar into the 'Hong Kong of the Mediterranean' all played a formative role in the party's electoral success.

With the Brussels Agreement providing the context for Anglo-Spanish discussions on Gibraltar's future status the annual meeting of the British and Spanish foreign ministers has afforded a forum within which negotiations can take place. Yet, it would appear that with the question of the utilisation of Gibraltar airport agreed between London and Madrid, neither Government is anxious that their growing rapprochement should be disturbed at this time by further consideration of an issue which has for so long bedevilled their accord. Other mutual concerns have again served to ensure that Gibraltar occupies a lowly position on their agreed agenda. Developments set in motion by the Soviet President, Mr Mikhail Gorbachev, with his doctrine of *glasnost*, and the resulting liberality which has been evidenced in the countries of the Warsaw Pact and the developing *détente* which has come to characterise relations between Moscow and Washington have meant that NATO is in the process of redefining its role and that Britain and Spain, as partners in the Organisation, have to take cognisance of such changes both individually and in their shared discourse. Likewise, the movement of the European Community towards the introduction of an Open Market in 1992 is an issue in which both Britain and Spain have a shared concern as they seek a format conducive to their common and discrete national interests. The growing interdependence occasioned by these broad and extensive changes had meant that neither London nor Madrid is currently prepared to contemplate the reawakening of such a divisive issue as Gibraltar especially when the Rock appears to be of diminishing importance in the grand scheme of their recast national priorities.

If Anglo-Spanish relations are being shaped in an ever-widening context and if, in consequence, Gibraltar has fallen from its former pre-eminence in that discourse it has to be conceded that since the Gibraltarian election of 1988 the Rock itself has actively sought to become the master of its own destiny as it has been increasingly freed from the constraints imposed upon it by the realpolitik of both its immediate neighbour and of its former colonial mistress. The Gibraltar Government has readily accepted the autonomy which has been bestowed upon it; even though the resulting flexibility has been more the gift of chance than of design.

Whilst it could be argued that the fundamental tenets of the socialism of the GSLP have been imported from Britain it has also to be accepted that they have not been adopted without modification to the requirements of Gibraltar's needs. The emphasis upon the attainment of socialist objectives through the pursuance of avowedly socialist means has exposed socialists in Britain to the charge that means have gained ascendency over ends and that it is through the pursuit of socialist means that socialist commitment may best be expressed in a capitalist economic system. In Gibraltar, the GSLP has emphasised the socialism encompassed within the ends sought rather than within the means used for their attainment. In consequence, it is possible to find Gibraltarian socialist ministers ready to praise the enterprise culture, which has been the objective of successive Conservative Governments in Britain since 1979, whilst contending that their brand of socialist thought is compatible with such an approach and departs from it in advocating that the distribution of the ends achieved should be on the widest possible communal basis and the allocation on grounds of greatest need. This has led the GSLP Government to actively court business partners and to establish joint venture companies in which representatives of private capital and government work together to revitalise the local economy. The role of an overtly socialist government has thus become that of a facilitator and overseer of developments undertaken in conjunction with private-sector enterprises and as a guardian of socialist objectives through seeking to ensure that the economic benefits of such collaboration fulfil the objective of collective advancement for the greatest possible number of its citizens.

Rather than pursuing policies dictated by socialist orthodoxy the GSLP has, since it came to office, sought to further the cause of effective democratic socialism. Hence the claim is made that socialist dogma has been replaced by socialism in action through the adoption of means congruent with the major social forces in Gibraltarian society whilst remaining compatible with deeply held socialist tenets.

Such an approach remains in its infancy and will finally be judged by its ability to secure the ends to which it is directed. Whether or not there is the political experience present within the ministerial ranks of the GSLP to carry through such a programme is not yet clear but there can be no doubt but that the political will to do so is abundantly self-evident. With the Spanish Government prepared to concede in private that the frontier restrictions imposed upon Gibraltar during the Franco era were both mistaken and counterproductive, but unwilling to relinquish its claim to sovereignty, and with the British Government, in public, continuing to leave the question of Gibraltar's future status to the freely and democratically expressed wishes of its citizens, the fundamental constraints acting upon the Rock's destiny appear to be as firmly in place as ever. However, the growing rapprochement between London and Madrid has brought the Rock's Government a breathing space and it has lost no opportunity to capitalise on this unanticipated but warmly welcomed respite. It has actively striven to develop an economic infra-structure capable of withstanding future shock and as a prerequisite for increased independence and has been encouraged in this by the recent resolution of the pensions dispute. Whether or not the Government of Gibraltar will be afforded the opportunity to continue to act as the master of its own destiny or whether it will again find itself constrained by the inability of Britain and Spain to agree on the future course that the Rock should take remains unclear but is certain to be one which will be determined beyond the narrow confines of the Rock and despite the best intentions of its government; no matter how great or worthy those may be. Proof of this was offered as 1990 drew towards its close, when, at a meeting of the Conference on Security and Cooperation in Europe, held in Paris in November, the Spanish Premier, Señor Felipe Gonzalez, made use of that platform to reiterate the Spanish claim to sovereignty over Gibraltar.

NOTES

1 THE AWAKENING PROBLEM

1 The pre-war Gibraltar City Council, whose powers and responsibilities were vested in the Governor during the war, had consisted of 5 officials and 4 elected members. Under the 1944 constitution the City Council was enlarged to 12 members (6 official and 6 elected) with an elected member as chairman. The first elections were scheduled for May 1945 and the electorate was to comprise all males over 21 years of age possessing the franchise. An Advisory Council was also created comprising the entire City Council, the Colonial Secretary and the Attorney-General.

Details of a further new constitution for Gibraltar, providing for a Legislative Council and an Executive Council, were given to the Governor of the Colony in a letter from the Colonial Secretary on 4 August 1949. A Legislative Council, presided over by the Governor, was to have 3 ex officio members (the Colonial Secretary, Attorney-General and Financial Secretary), 5 elected members and 2 nominated members (of whom both could be, and one must be, official). Powers were reserved to the Governor to deal, in default of action by the Legislative Council, with matters relating to defence, public order and good government. The Governor also had the power to veto Bills and to initiate legislation held to be necessary. An Executive Council was to be largely responsible for the framing of the legislative programme, although at least one, and possibly more, of its three unofficial members would be drawn from the Legislative Council.

2 Lord Tryon, First Commissioner for Works, told the House of Lords on 13 August 1940, that accommodation had been provided, in London, for 11,000 evacuees from Gibraltar.

A written parliamentary answer in the House of Commons on 18 December 1940, stated that 'refugees' from Gibraltar, in the United Kingdom, number 5,359 women, 3,920 children and 1,690 men, all of whom were accommodated in London and that it had been intended to transfer them to a more congenial climate before the on-set of winter but this had not been possible.

3 Conditions in the reception camps had been the subject of numerous

and vociferous complaints, and demonstrations calling for repatriation 'continued until the end of 1947, when some Gibraltarians had still not returned home. The process was not completed until 1951'. Dennis, Philip *Gibraltar*, David and Charles, Newton Abbot, 1977, p. 95.

4 See Northedge, F.S. *Descent from Power: British Foreign Policy 1945–1973*, George Allen and Unwin, London, 1974.

5 Dennis, *op. cit.*, p. 95.

6 See Haigh, R.H., Morris, D.S., and Peters, A.R. *The Guardian Book of the Spanish Civil War*, Wildwood House, London, 1987.

7 Medhurst, Kenneth N. *Government in Spain: The Executive at Work*, Pergamon Press, Oxford, 1973, pp. 23–4. Spain was, for example, excluded from the Marshall Aid Programme and denied membership of the United Nations until 1955.

For a detailed coverage of Spanish foreign policy during the Second World War, see, Gallo, Max *Spain Under Franco*, George Allen and Unwin, London, 1973, and Hills, George, *Franco: The Man and His Nation*, Robert Hale, London, 1967.

8 Ibid., p. 24.

9 Ibid., p. 25.

10 Ibid.

11 These agreements were the outcome of some eighteen months of negotiations after informal discussions had been initiated by US Admiral Sherman in July 1951.

12 *New York Times*, 28 September 1953.

13 The Portuguese Minister of Foreign Affairs, Professor Paulo Cunha, echoed Franco's views in an interview with *The Diario de Noticias* on 1 October 1953: 'The Portuguese Government's attitude regarding Spain and Western Europe was clearly defined long ago. It is with true satisfaction that we see this important step taken for the inclusion of Spain in the western defence system. The US–Spanish treaty does not clash with . . . the 1939 treaty . . . between Portugal and Spain . . . it strengthens the inviolability of the metropolitan territories of the two countries.'

14 *The Times*, 28 September 1953.

15 Medhurst, *op. cit.*, p. 25.

16 Ibid.

17 The visit took place between 20 August and 3 September 1959 and in part coincided with a visit to Britain being made by US President Eisenhower with whom Señor Castiella met on 31 August.

18 The visit took place on 11–13 July 1960.

19 The visit to Portugal took place on 26–9 May, and to Spain on 29–31 May 1961.

20 Questions were raised in the House of Commons on 30 May 1961.

21 Although the closing communiqué made no mention of what was discussed, Lord Home said at a press conference, on 31 May, that there had been joint consideration of the 'mobilisation of the economic defence of Europe' and 'the general world picture'. The two foreign ministers also signed an agreement supplementing the Consular provisions contained in the Anglo-Spanish Treaty of Commerce and

Navigation of 1922, and exchanged ratification documents of the Anglo-Spanish cultural convention signed in London during Señor Castiella's official visit in the previous year.

2 THE PROBLEM GROWS

1 The new constitution agreed for Gibraltar on 10 April 1964 came into effect in August 1964 with the first elections to the new Legislative Council being scheduled for September of that year was as follows:

(a) The number of elected members of the Legislative Council sitting on the Executive Committee would be increased from 4 to 5, and the Executive Council would in future be known as the Gibraltar Council.

(b) The Chief Member would in future be known as the Chief Minister. He would be the person whom the Governor considered to command the greatest measure of confidence among the elected members.

(c) The Governor would apportion departmental responsibilities after consultation with the Chief Minister. Any elected member of the Legislative Council given departmental responsibilities would be styled Minister and would be responsible for his department. All Ministers would be collectively responsible for decisions of the Council of Ministers or of the Gibraltar Council with respect to matters assigned to Ministers.

(d) The Chief Minister would be Leader of the House and vested with the direction of Government business.

(e) As the Chief Minister would be in charge of Government business, the Chief Secretary would have few duties to perform in the House and would cease to be a member of the Legislature; he would, in future, be known as the Permanent Secretary to the Government of Gibraltar.

(f) The number of elected members of the Legislative Council would be increased from 7 to 11, giving the Governor a larger field from which to select Ministers.

(g) There would no longer be any nominated members in the Legislative Council.

(h) Responsibility for the appointment of elected members to the Gibraltar Council would rest with the Governor, after consultation with the Chief Minister.

(i) In place of the present Council of Ministers, there would be a Council of Ministers presided over by the Chief Minister, consisting of the other elected members of the Gibraltar Council and such other Ministers as the Chief Minister might designate as members of the Council. Matters within the responsibility of Ministers would normally come direct to the Council of Ministers. As a general rule, the recommendations of the Council of Ministers on matters of purely domestic concern would be endorsed by the Governor in Council.

In view of the adoption of a resolution by the Legislative Council following a merger between the City Council and the Government, and of assurances that the Government was working out detailed proposals, it was agreed that the future of the City Council should be decided early in the life of the next Legislature.

2 *Daily Express*, 9 June 1964. It is not possible to be precise about the specific details of the naval construction programme but it would seem to have entailed the building, in Spanish shipyards of: a light cruiser, 4 frigates and 2 submarines using British licences, materials and technicians.

3 In a written reply on 7 July 1964, the Minister of Defence, Mr Peter Thorneycroft, indicated that the naval construction programme would have been worth some £11,000,000 to the British economy – £8,000,000 to £9,000,000 for equipment manufactured in Britain for the Spanish Government, and £2,000,000 to £3,000,000 in design fees.

4 British sovereignty of Gibraltar stems from Article 10 of the Treaty of Utrecht (1713) which provided that the session of Gibraltar by Spain to Britain was absolute and 'to be held and enjoyed absolutely with all manner of right for ever, without any exception or impediment whatsoever'. The Treaty of Utrecht was reaffirmed in the Treaties of Paris (1763) and Versailles (1783).

5 See UN General Assembly Resolution 1514, 14 December 1960, on ending colonialism.

6 Representatives of Venezuela, Uruguay and Mali also spoke in the UN debate on Gibraltar and all called for Britain to negotiate on a solution to the problem with Spain. Australia lent support to the British and Gibraltarian case whilst the USSR demanded the de-militarisation of Gibraltar and alleged that Spain had the intention of converting the Rock into a joint Spanish–British military base.

7 Some one thousand British subjects from Gibraltar, the UK and India and some Portuguese citizens were affected by the restrictions on passports and workers' passes; the majority of these had to leave their homes in Spain and were provided with temporary accommodation in Gibraltar.

8 This constitutional change had been effected in 1950. See above.

9 At this time the Senior Economic Adviser to the Colonial Office, Mr Percy Selwyn, was in Gibraltar for consultation with a view to presenting a report to the Gibraltar Government.

10 Mrs White was in Gibraltar from 12–15 February 1965.

11 The withdrawal of workers' passes from British subjects was made effective from 7 March 1965.

12 See above.

13 *Gibraltar – Recent Differences With Spain*, Cmnd 2632, HMSO, London, 1965.

14 The 'master plan' was a longer-term project. More pragmatic and immediately utilisable proposals had been announced on 3 February 1965 when the Gibraltar Government had begun negotiations with Morocco, Portugal and Malta for the recruitment of contract workers for

the dockyard and building industry. This initiative had been a response to fears that the Spanish Government would ultimately completely block the daily entry of Spanish workers into Gibraltar; estimated at approximately 10,000.

15 Sir Joshua Hassan left London on 27 July after extensive talks with British Ministers on a range of matters relating to the Colony. He was accompanied on his visit by Mr Peter Isola who had, since 8 July been Deputy Chief Minister: a date which marked the formation of a coalition government in Gibraltar; Mr Isola having been, until that date, Leader of the Opposition.

3 A DEEPENING PROBLEM

1 This resolution was supported in the General Assembly by 96 votes to 0, with 11 abstentions. Both Britain and Spain voted for the resolution. France, Portugal, the Soviet Union and 8 other Communist countries abstained.

2 This clearly implied a diminution in Gibraltar's self-government.

3 There was no suggestion that the Spanish representative would participate in the Government of Gibraltar, which would remain in the hands of the Governor and the elected representatives.

4 Smuggling had been an issue made much of in the Spanish media as a reason for the initial imposition of frontier restrictions and had also been cited by Government officials.

5 The third round of talks were held on 6–7 September 1966.

6 This took effect from 4 August 1966.

7 The UN resolution referred to in the Spanish Note of 14 December had been adopted, on 17 November by the Special Committee on Colonialism (Committee of 24). The voting had been 16 in favour and none against with Britain, USA, Australia, USSR, Poland and Bulgaria abstaining and two countries absent.

8 On 15 October, in response to Spanish allegations that there had been 12 violations of Spanish airspace by RAF planes in the Gibraltar area, the British Embassy in Madrid stated that there was no evidence that 9 of the alleged violations had been committed, that in 2 cases there was no conclusive evidence, and that in 1 case an RAF plane had inadvertently flown over Spanish territory, for which an apology was tendered. On 22 October, the Spanish Government presented another Note alleging a further 15 violations of Spanish airspace in the Gibraltar area, and asking the British Government to reconsider their declarations of sovereignty over the isthmus and to stop using the Gibraltar airstrip for military purposes. The following day, the British Foreign Office stated that the British Government had 'no doubt about British Sovereignty over the ground on which Gibraltar airport stands, which has long been administered as British sovereign territory'; that the British Government also had no doubt as to 'the right to use the airfield for military purposes as they had done for 25 years'; and that the British Government intended to continue using the airfield for these purposes. Allegations of further infringements provided the content of a

further Spanish Note on 30 November. The British Government replied on 5 January 1967, denying the alleged infringements of Spanish airspace and warning that the Spanish Government would be held responsible for the consequences of any actions Spain might take 'allegedly in defence of its sovereignty and territorial integrity' or 'involving any infringment of British rights'.

9 The Spanish Foreign Ministry issued a statement dissociating itself from this action and stated that violence could only obstruct the negotiations with Britain over Gibraltar, and apologising for the incident.

10 The resolution passed by the Trusteeship Committee and the General Assembly accorded almost exactly with an earlier resolution approved by the Committee of 24 at its meeting of 17 November. There was, however, one significant difference, namely, the insertion into the later resolution of the phrase in (ii) 'taking into account the interests of the people of the territory'.

11 Sir Joshua Hassan had visited London before travelling on to New York, for talks with Mr Lee during the course of which he stated that he intended to explain to the UN the determination of the Gibraltarians not to become part of Spain nor to come under Spanish sovereignty.

12 Under ICAO regulations, Spain was entitled to declare prohibited areas on the grounds of military necessity or public safety. However, the regulations demanded that the prohibited areas should not cause unnecessary interference to aerial navigation. It was the British contention that the area defined by the Spanish Government did occasion such unnecessary interference and was, therefore, contrary to ICAO regulations.

13 The following day, Mr Frederick Mulley, Minister of State for Foreign Affairs, met with the Spanish Ambassador, Marques de Santa Cruz, and reiterated that Britain would maintain its rights in Gibraltar and that the airfield there would continue to be used for civilian and military aircraft. He emphasised that it was the action of the Spanish Government that had directly led to the postponement of the Anglo-Spanish talks scheduled for 18 April.

14 The British Note stated that advanced radar equipment was being used at Gibraltar to prevent any violations of Spanish airspace, that 2,500 landings or take-offs had been made at Gibraltar since May 1966 and that there had been 'only one clear instance of an infringement of Spanish rights arising directly out of the use of Gibraltar airfield by British military aircraft' which had been by a trainee officer. The British Note conceded two inadvertent violations: one arising from an administrative error, and another stemming from a navigational error because of heavy cloud in the Straits. It was maintained by the British Government that standard approach procedures necessary for access to Gibraltar airfield 'involves no infringement of Spanish rights'.

15 Under ICAO regulations Spain was entitled to declare prohibited areas on the grounds of military necessity or public safety; the same regulations laid down that the area in question had to be of reasonable extent and location and should not interfere unnecessarily with aerial navigation. In the British view the area defined by Spain was of

unreasonable extent and location, and the prohibition was completely unjustified and a violation of the Chicago Convention. Britain asked the Council of the ICAO to consider the dispute with Spain under Article 84 of the Chicago Convention, which related specifically to disputes between member nations of the ICAO.

16 The first British civilian aircraft to land at Gibraltar airfield after the imposition of the Spanish ban was a BEA Comet with 57 passengers on board which touched down at 0800 hours on 15 May. Two Spanish jet fighters patrolled the perimeter of the restricted zone as the Comet approached the airfield. Subsequently, other British planes landed unhindered at Gibraltar, all having avoided Spanish airspace on the last stage of the flight. On the first day of restrictions RAF Hunter jet fighters patrolled around the Rock.

17 The team was to be led by Mr R.L. Hutchens, New Zealand Ambassador in Paris; the other members were to be Mr Daniel Owino, Kenyan Ambassador in Bonn; Dr Kenneth Rattray, Assistant Attorney-General in Jamaica; and Mr Mustansir Rahmann, of the Commonwealth Secretariat in London.

18 Presented by Chile, Iraq, Syria and Uruguay the resolution was supported by 16 votes to 2 with 6 abstentions. Britain and Australia voted against. The resolution: (1) deplored the interruption of Anglo-Spanish negotiations on Gibraltar; (2) declared that the holding of a referendum by Britain contradicted previous UN resolutions; and (3) invited the British and Spanish Governments to resume negotiations without delay and with a view to 'ending the colonial situation in Gibraltar and safeguarding the interests of the population when this colonial situation shall have ended'.

4 NO SOLUTION IN SIGHT

1 The resolution was passed by 70 votes to 21 with 21 abstentions. The resolution was sponsored by Argentina, Colombia, Ecuador, Honduras and Panama.

2 The resolution was supported by all the Communist countries, with the exception of Albania, all the Arab countries, all the Latin American countries, except Mexico, a majority of the Asian countries, many African countries and, among the European countries, by Greece, Ireland, Italy, Portugal and, inevitably, by Spain. The countries voting against the resolution, in addition, equally inevitably, to Britain were: Australia, Barbados, Botswana, Canada, Ceylon, Denmark, Gambia, Guyana, Jamaica, Malawi, Malaysia, Maldive Islands, Malta, New Zealand, Norway, Sierra Leone, Sweden and Trinidad and Tobago. Abstentions were recorded by: Austria, Belgium, Central African Republic, Congo, Cyprus, Ethiopia, Finland, France, Ghana, Iceland, India, Israel, Kenya, Laos, Luxemburg, Madagascar, Mexico, Nepal, Netherlands, Niger, Nigeria, Senegal, Singapore, Thailand, Togo, Uganda and the USA. The absentees were: Albania, Kuwait, Lesotho and South Africa.

3 ABC 20 December 1967.

4 *Gibraltar Chronicle*, 4 April 1968.

5 We have ascertained the names of the members of the group known as the 'Doves' and it seems reasonable to contend that their proposals emanated from the worsening economic situation developing in Gibraltar; a factor emanating directly from Spanish frontier restrictions.

6 In the course of the demonstrations, 12 policemen were injured and 16 protestors arrested.

7 The option clause of the Treaty of Utrecht provided that, in the event of a surrender of British sovereignty, Gibraltar should first be offered to Spain.

8 The following seventeen countries voted with Britain in opposing the General Assembly's resolution: Australia, Barbados, Botswana, Canada, Denmark, Guyana, Jamaica, Lesotho, Liberia, Malawi, Malaysia, Maldive Islands, Mauritius, New Zealand, Sierra Leone, Singapore and Sweden. Abstentions were registered by the United States, France, Finland, Iceland, Ireland, Italy, Japan, Netherlands and Norway. It is interesting to compare this voting with that at the 22nd session of the General Assembly the previous year: see Notes to Chapter 3.

5 CONTINUING GLOOM

1 The constitution stipulated that Gibraltar would no longer be a colony but would, henceforth, be termed the 'City of Gibraltar'.

2 Many of Gibraltar's Spanish workforce had never worked elsewhere and many of them had been employed in Gibraltar for over twenty years.

3 The ordinance of 10 June 1969 was valid for 6 months. The right of appeal to a tribunal was, however, contained within the ordinance.

4 *The Times*, 9 June 1969.

5 This meant that Gibraltar's only remaining link to the Spanish mainland was through the twice-weekly air service operated by British Airways to London via Madrid.

6 Under the decree, Gibraltarians accepting Spanish nationality would be allowed to take into Spain household goods, including personal belongings, furniture, cars and small boats, free of customs duty and without payment of taxes. Likewise, businesses, factories and shops could be transferred to Spain without payment of taxes. Recognition was extended to professional qualifications, thus permitting teachers, doctors and lawyers to practise in Spain. Gibraltarians would have the option of retaining their British nationality should Gibraltar pass under Spanish sovereignty, in which case they would be afforded complete autonomy in such internal spheres as police courts, the economy and local government.

7 *The Times*, 6 July 1969.

8 The Gibraltar Labour Party had previously been known as the Association for the Advancement of Civil Liberties (AACL). Mr Peter Isola had previously been with the AACL. The turn-out was 71 per cent at the first general election under the new constitution.

9 Statement to the House of Commons, 13 October 1969.

10 On 28 September, the Spanish newspaper, *Ya*, called for a sea blockade of Gibraltar. The following day, four Spanish warships, two corvettes

and two minesweepers, anchored off the northern end of Gibraltar's airport to be joined, on 8 October, by the cruiser *Canarias* and the helicopter carrier *Dedalo*. Four Spanish destroyers also made a brief appearance in Algeciras Bay at a distance of a quarter of a mile off Gibraltar. On 2 October the British Government responded to the Spanish naval presence and ten British warships including the carrier *Eagle*, the commando carrier *Bulwark* and the guided missile destroyer *Hampshire* had added to the naval congestion in the waters around the Rock.

11 Whether 'manoeuvres' refers to military and/or political actions by the British Government is unclear.

12 This letter from Lord Caradon to U Thant was published on 9 October 1969.

13 The announcement was made on 8 December 1969. Under the development programme were to be included: a start on 750 housing units on land to be released by the British Ministry of Defence; the building of hostel accommodation for temporary immigrants arriving (primarily from Morocco) in Gibraltar to take over jobs in the shipyard and hotels formerly filled by Spaniards; the provision of a comprehensive secondary school system; medical and recreational facilities; and tourist development.

14 Sir John Russell met with General Franco on 6 November 1969.

15 Interview with Señor Lopez Bravo in the Spanish newspaper, *ABC*, 18 December 1969.

16 The Spanish newspaper, *Informaciones*, reported private conversations between Señor Lopez Bravo and Sir John Russell, 11 May 1970.

17 Confusingly, the British Foreign Office confirmed, on the same day, reports in the newspaper, *Informaciones*, that informal talks between the Spanish and British Government had been continuing for some 7 months with the overriding aim, on the British side, of removing restrictions on Gibraltar.

18 The Minister, however, remained adamant as to the Spanish claim to sovereignty over Gibraltar, 'Nothing will be done that might lessen . . . or cast the slightest doubt on our firm intention to restore the territorial integrity of Spain by incorporating into our national sovereignty the Rock.'

19 The preparedness of the EC to consider any links between Britain and the EC as applicable to Gibraltar was given wide publicity in the Spanish press because it neglected Spanish claims to sovereignty over the Rock. On 17 November 1970 the Spanish Government informed the EC of its concern over the prospect of Gibraltar being associated with the Community upon Britain's entry. The Spanish Government also emphasised that such a decision would be contrary to resolutions passed by the UN on Gibraltar.

20 As evidence, Señor Bravo cited recently released British Government Papers in which reference was made to a plan, prepared in 1940, to blow up Spanish posts in the event of Spain joining the Axis powers. The explosives necessary to the execution of that plan had been stored at Gibraltar.

21 Sir Denis Greenwood visited Madrid on 3 June 1971. The official communiqué described the talks as having taken place in an 'amicable atmosphere'.

22 Sir Alec Douglas-Home visited Madrid on 27 February – 1 March 1972. His arrival was greeted by demonstrations staged by young, right-wing extremists who sang the Falange anthem, gave the fascist salute and called for Britain to leave Gibraltar. This was the first visit by a British Foreign Secretary to Spain since Sir Alec himself went there in 1961.

23 Interview with Señor Lopez Bravo in the Spanish newspaper *Hoja del Lunes*, 18 October 1971.

24 *The Times*, 2 March 1972.

25 See the *Gibraltar Post*'s coverage of this election.

26 The TGWU has long been a powerful force in Gibraltarian politics and enjoys what amounts to a monopoly position in the representation of labour on the Rock.

27 The question of Sir Joshua's role as a 'broker' between Britain and Spain was to re-emerge in 1974.

28 *Financial Times*, 6 April 1973.

29 Señor Lopez Bravo's failure to return Gibraltar to Spanish sovereignty appears to have caused his dismissal. His softer line on Gibraltar had stood in marked contrast to that of his predecessor but his inability to bring Britain to the negotiating table, despite his seven meetings with the British Foreign Secretary, had tried the patience of the right wing of his party which subjected him to hostile criticism in the Cortes.

30 Spanish communication to the UN Secretary-General, 18 July 1973.

6 A GLIMMER OF HOPE

1 The British delegation had contended that the restrictions imposed by Spain on aircraft movements in the Algeciras region presented a real threat, in certain weather conditions, to the safety of flights to and from Gibraltar; particularly to the wider-bodied aircraft that were increasingly coming into use by civilian airlines. As the Madrid talks ended the Spanish Cabinet discussed plans to build a large commercial airport at Castellar de la Frontera, some 9 miles from Gibraltar. Had such a scheme gone ahead it would have created serious air traffic control problems for the Gibraltar airport which is situated across the narrow neck of the isthmus which joins the Rock to the Spanish mainland.

2 Señor Juan Roldan was the chief London correspondent of the Spanish news agency EFE. His article, entitled 'A Spanish View; An Open Wound in the Heart', appeared in *The Times*, 30 September 1974.

3 Sir Joshua Hassan's letter appeared in *The Times* of 14 October 1974.

4 See *The Times*, 23 October 1974.

5 See Chapter 5.

6 *The Times*, 7 November 1974.

7 Sir Alec Douglas-Home was Secretary of State at the Foreign and Commonwealth Office in London at the time of Sir Joshua's meeting in Brussels with the representative of the Spanish Foreign Ministry.

8 See *The Times*, 12 November 1974.

9 Señor Lopez Bravo visited London in July 1972.

10 Mr Xiberras found it 'ironic' that it was through public correspondence between Gibraltar's Chief Minister and the Spanish Ambassador in London that the people of the Rock should learn of the Brussels meeting and the Spanish proposals. He did, however, welcome the fact that the Spanish proposals had been made public 'because the folly of entrusting Spain with the sovereignty of Gibraltar, and with it the completely unworkable and unacceptable legal, political and human consequences of such an act . . . cannot fail to escape the notice of even that handful of Gibraltarians who imagined than an agreement with Spain is possible with only nominal concessions on sovereignty'.

11 See *The Times*, 12 November 1974.

12 See *The Times*, 30 September 1974.

13 Sir Joshua Hassan, Sir John Grandy and the Gibraltar delegation were in London on 4–12 November 1974.

14 New projects stipulated in the three-year aid programme included £3,450,000 for housing, £1,878,000 for education and £1,330,000 for expenditure on electricity generating plant, port development and hospital renovation. Provision was also made for the Ministry of Overseas Development to consider the addition of other new projects to the aid programme when the results of detailed reviews were available. These reviews of progress were to be undertaken at the end of the second year of the three year programme. In April 1974 agreement had been reached between the British and Gibraltar Governments on the transfer to civil use of land no longer needed for defence purposes.

15 *La Vanguardia*, 24 October 1974.

16 This review took place in September 1974.

17 The general strike of 1972 had lasted from 20–26 August. The Gibraltar Trades Council had then accepted the employers' offer of a minimum weekly wage of £14 for manual workers and £17 for white-collar workers. It should be noted that a high percentage of the Rock's workforce were employed in the public sector either as employees of British or Gibraltarian Government departments. Since 1972, average weekly earnings had risen to approximately £25 through the automatic payment of cost-of-living allowances linked to the retail price index and also because of the opportunities for dual employment and overtime working arising from the labour shortage in Gibraltar; a shortage stemming from the loss of the former Spanish workforce with the closure of the frontier in 1969.

18 Available evidence would suggest that the Gibraltar branch of the Transport and General Workers' Union was closely identified with the campaign conducted by the Integration With Britain Party and hence was opposed to many of the policies of the governing Gibraltar Labour Party.

19 By October 1976, some two years after the beginning of the parity dispute, the adoption of the Scamp recommendations had produced an increase in the basic wage of labourers, the lowest paid category, from £17.50 to £25.00 per week. During the same period the salary of a qualified teacher had risen from £1,700 per annum to £2,615 per annum.

20 Mr Hattersley was in Gibraltar on 24–26 September 1975.

21 Sir John Grandy's reply to Mr Xiberra's letter to him was dated 13 October 1975.

22 See Mr Maurice Xiberras's letter to *The Times* of 28 October 1975.

23 The telephone lines were re-opened from midday on 24 December 1975 until midnight on 2 January 1976. Approximately 1000 phone calls per day were made between Gibraltar and the Spanish mainland during this period. Telephonic communication was also restored over Easter; from 10–20 April 1976.

24 Señor Areilza was a diplomat of long-standing who had enjoyed considerable prestige under the Franco regime until, from roughly 1966, he began to engage in open criticism of that regime.

25 Talks were held in London on 24–5 June 1976 between Sir Joshua Hassan, Chief Minister, Mr Maurice Xiberras, Leader of the Opposition, and Mr Roy Hattersley, Minister of State at the Foreign and Commonwealth Office. Plans for constitutional reforms for Gibraltar which involved greater integration with Britain constituted the agenda. The proposed reforms would have modified the judicial status of the Rock and strengthened its ties with Britain and would have also created a permanent economic link. The proposal was rejected in a formal memorandum issued after the talks, on the grounds that 'innovations which might hamper the development of a more favourable Spanish attitude' should be avoided. The memorandum emphasised that without a change in Spanish attitudes there could not be a long-term solution to Gibraltar's problems especially as these emanated from the 'restrictive measures' which had been progressively enforced by Spain since 1966. The proposed reforms were the recommendations of the cross-party Gibraltar Constitutional Committee referred to above.

26 Mr Xiberras was later to write that the British Government had 'deliberately destroyed the Integration With Britain Party'. See his letter to *The Times* of 6 October 1976.

27 Of these three candidates, Mr Jose Emmanuel Triay, a lawyer, was the chief proponent of close links with Spain and a Gibraltarian presence at Anglo-Spanish negotiations. He had gained a measure of political notoriety as a member of The 'Doves': a group which had advocated a view markedly similar to that which he now advanced in 1968. (see above). Mr Triay's political stance was shared by Mr Benady whilst, the third member of the trio, Mr Eric Ellul favoured an economic, commercial and cultural agreement with Spain which he envisaged evolving into a political agreement over time.

28 The label 'Gibraltar Labour Party' which had featured prominently in the party's 1972 election campaign was noticeably played down in 1976 and prominence was again given to the label under which the party had campaigned so successfully since 1945, namely, the 'Association for the Advancement of Civil Rights'.

7 HOPE IS SUCCOURED

1 Señor Oreja's speech in San Roque, a town overlooking Gibraltar, was

delivered on 6 February 1977.

2 *Gibraltar: British or Spanish? The Economic Prospects*, Witton House, London, 1977. The collaborative survey had been undertaken over a 3-year period by two teams of consultants, one British, Maxwell Stamp Associates, and one Spanish, Iberplan, and had been financed by business interests and private foundations. It was stressed that the conclusions had been reached independently of the British, Spanish and Gibraltar Governments.

3 This was the first Spanish Cabinet meeting since the general election of 15 June. Among the economic matters discussed were the impending devaluation of the peseta, tax reform, measures to stimulate the economy and reduce unemployment, a more flexible credit structure and other forms of assistance for small and medium-sized businesses. Interestingly, the question of Gibraltar had been ignored by all parties in the campaign leading to the election.

4 Greece had sought full membership of the EC on 12 June 1975 and Portugal had done likewise on 28 March 1977. Both Spain and Portugal had already been linked to the EC through preferential trade agreements which had been effective from 1 October 1970 and 1 January 1977.

5 Dr David Owen was in Madrid on 5–7 September 1977.

6 The Party for the Autonomy of Gibraltar had been founded on 7 September 1977 by Mr Triay. A member of the group known as the 'Doves', Mr Triay had stood as an Independent at the last general election in Gibraltar.

7 Señor Oreja addressed the General Assembly of the UN on 26 September 1977.

8 The foreign policy debate in the Cortes took place on 20 September 1977.

9 The 'Committee for Reconciliation' had been formed by Señor Gonzalez Arias Bonet, a self-proclaimed pacifist. The members of the Committee supported the Spanish Government's claim to sovereignty over Gibraltar but were opposed to the means being used, such as closure of the border. The demonstration lasted for two hours before coming to a peaceful end.

10 Señor Suarez visited London on 19 October 1977.

11 The visiting party comprised the Gibraltar Chief Minister, Sir Joshua Hassan, the Leader of the Opposition in the House of Assembly (The Independent Deputy and former Leader of the Integration With Britain Party), Mr Maurice Xiberras and the Governor of Gibraltar, Sir John Grandy. The party arrived in London on 3 November 1977.

12 Sir Joshua Hassan's statement to the House of Assembly was made on 8 November 1977.

13 In Strasburg, on 24 November 1977, Señor Oreja deposited documents of accession for Spanish membership of the Council of Europe and also signed the European Convention for the Protection of Human Rights and Fundamental Freedoms. (How this latter act squared with the hardship still being caused to the inhabitants of Gibraltar by Spanish frontier restrictions is a matter of diplomatic nicety.) Spain's application to join the Council of Europe had been unanimously approved by the Council's Parliamentary Assembly on 12 October 1977, its unanimous

ratification by the Spanish Congress of Deputies and Senate followed on 17 and 18 November respectively. The application was accepted by the Council's Committee of Ministers immediately prior to the formal admission ceremony. Spain thus became the twentieth member of the Council of Europe.

14 The 'new spirit' was the term coined by Señor Oreja at Strasburg to describe the increasing amicability of Anglo-Spanish relations.

15 Señor Oreja addressed the Spanish Parliament on 11 January 1978.

16 Señor Ruperez departed for Gibraltar on 24 January 1978.

17 Señor Oreja's address to foreign correspondents in Madrid was given on 30 January 1978.

18 The Canary Islands Independence Movement (MPBIAC) was led by Señor Antonio Cubillo and was based in Algeria and, as its name implied, was seeking independence for the Canaries from Spanish rule. The Canaries had been ruled by Spain for centuries and the inhabitants were of Spanish descent. Geographically, the Canaries lay closer to Africa than to the Spanish mainland and it was to protect Spanish rule in the Islands that Spain, in the nineteenth century, established a Saharan Colony on the African coast from which it eventually withdrew in 1976.

19 Not surprisingly, Morocco and Mauritania were the only members of the OAU to vote against the resolution.

20 A counter-ruling to that produced by the OAU on the sovereignty of the Canary Islands was forthcoming from the Council of Europe on 18 October 1979, when, at a conference organised for the members of local authorities and regional bodies of its 21 member nations, a resolution proposed by Dr Mota Amaral, President of the Azores Regional Government, affirming the European nature of the Canary Islands, the Azores and Madeira, was carried.

21 Morocco had agreed to put its claim to Ceuta and Melilla into abeyance in 1976 in return for a tripartite agreement, proposed by the last Franco Government, which covered the division of the former Spanish Colony of Spanish Sahara between Morocco and Mauritania. The parallel with Gibraltar was drawn by the Madrid newspaper *Diario-16* in a leading article of 15 March 1978.

22 Spain had had defence agreements with the USA since 1953 and the agreement on American military bases in Spain had last been renewed in 1976 ('Treaty of Friendship and Co-operation between Spain and the USA'; signed 24 January 1976) and was scheduled to remain in force until 1981. This Treaty replaced the 1970 'Agreement of Friendship and Co-operation' which officially expired on 26 September 1975. Prior to 1976 all USA–Spanish Agreements had been regarded in Washington as being exclusively 'Executive agreements' and thus did not require the approval of the American Congress. Concern had been voiced in the US Senate after the conclusion of the 1970 Agreement that such an important foreign policy document had not been put before it in the guise of a formal treaty. The negotiations of 1976, therefore, saw the light of day not as an executive agreement but as a treaty proper which was submitted to the Senate on 18 February 1976 and approved by its Foreign Relations Committee on 18 May 1976, by 11 votes to 2,

and received final ratification by the Senate on 21 June 1976 by 84 votes to 11. Three days prior to final ratification the Senate had authorised an initial $36 million in military aid to Spain for the first year of the Treaty. The Treaty was submitted to the Spanish Cortes on 12 March 1976 and secured the unanimous approval of its Foreign Affairs Committee on 23 June 1976.

23 In 1978 the numerical strength of Spanish military forces was, approximately, as follows: Army, 220,000; Navy, 47,000; Air Force, 34,000.

24 Sir Joshua did, however, stress that he found it perfectly proper that the British Government had not used the threat of withholding its support for Spanish membership of the EC as a means of exerting pressure upon Madrid to remove the frontier restrictions it had placed upon Gibraltar. He also made mention of the Spanish suggestion that Gibraltar could enjoy enhanced regional status upon passing into Spanish sovereignty but he envisaged such a prospect lying far in the future and as one which could only be contemplated after Spain had proved itself to be a good neighbour over many years.

25 After a visit to Gibraltar on 4–6 April 1978, Mrs Judith Hart, Minister of State for Overseas Development, had announced that British aid to Gibraltar for the period April 1978 to March 1981 would amount to £14 million.

26 President Ford and Secretary of State Kissinger were in Madrid on 31 May – 1 June 1975. The North Atlantic Council meeting in Brussels took place on 29–30 May 1975.

27 The three working parties had met in London on 17 July 1979. They were to meet again, in Madrid, on 14–15 December 1979. A Gibraltarian delegation was present at the Madrid meeting.

28 It should be noted that Greece, Portugal and Spain, the latest group of countries seeking membership of the EC, had large and relatively backward agricultural sectors which nevertheless produced agricultural products cheaply because of low overheads. In contrast, the earlier members of the EC had strength in the manufacturing sector. The agricultural policy of the EC, with its large budget, subsidies and the problem of surplus production, had long been a focus for heated debate among the member nations.

29 Lord Carrington's address to the House of Lords was on 26 June 1979.

30 These remarks were made in an interview reported in the local Gibraltar newspaper, *Panorama*, of 15 July 1979. These were the most overtly political remarks, it might almost be said that they were the most political partisan remarks, made by a Governor of Gibraltar on the Rock's future. General Sir William Jackson had become Governor on 30 May 1978 and had quickly developed a rapport with the city's people.

31 Support for the British Government's refusal to link Spanish accession to the EC to the lifting of Spanish frontier restrictions on Gibraltar was forthcoming from Conservative peers: Lord Selsdon contended that a solution to the Gibraltar question had to be related to Britain and Spain as members of the EC and Lord Morris, having claimed that the transition made by Spain from dictatorship to democracy had been one of the major achievements of the twentieth century, supported

the Government view that the lifting of Spanish sanctions on Gibraltar should not be made a pre-condition of Spanish membership of the EC.

32 This, of course, was not the first protest staged by Spaniards from the towns of the Campo region calling for the re-opening of the frontier with Gibraltar; nor was it to be the last.

33 There was talk in La Linea of further strike action and peaceful demonstrations in the future if Madrid failed to re-open the frontier.

34 This was not the first occasion on which the PSOE had called upon the Spanish Government to remove the frontier restrictions.

35 Presumably a 'uniqueness' which was out-moded and out of place in a world of ever-increasing international interdependence.

36 The seminar, the first of its kind, was held in Segovia on 7–8 December 1979. It was attended by non-governmental representatives from Spain, Gibraltar and Britain. Among those present was the former British Ambassador to Spain (1969–74) Sir John Russell, the Mayors of San Roque and La Linea, the two Spanish towns closest to Gibraltar and which, along with the Rock itself, had suffered most from the closure of the frontier, and a number of Spanish parliamentarians.

8 SIGNS OF PROGRESS

1 The incorporation of the notion of autonomy into the Party's title had been done to give recognition to the regional administrations which were being developed in Spain. The neighbouring Spanish region of Andalucia was to hold its referendum on autonomy in February 1981.

2 The DPGB's leader, Mr Peter Isola, had a long-lasting history of sitting on the Opposition benches in the House of Assembly. During 1979, Mr Isola had become Leader of the Opposition.

3 Two of those who had crossed the floor of the House, Mr Brian Perez and Dr Reginald Valarino, had joined the governing AACR–GLP led by Sir Joshua Hassan.

4 Collectively, the 'big four' political parties nominated 25 candidates: AACR–GLP, 8; DPGB, 8; GSLP, 6; PAG, 3. The two remaining candidates stood, unsuccessfully, as Independents. Some 10,600 Gibraltarians went to the polls on 6 February 1980.

5 The debate and vote in the Congress of Deputies took place on 27 March 1980.

6 Both Lord Carrington and Señor Oreja were in Lisbon for a meeting of the Council of Europe which took place on 6 April 1980. They conducted their talks in the Portuguese Foreign Ministry on 9–10 April.

7 This joint statement was issued on 12 April 1980.

8 Sir Joshua Hassan and Peter Isola departed for London on 13 April 1980.

9 Lord Carrington met for talks with the Gibraltarian representatives and told the House of Lords of the Lisbon Agreement on 14 April 1980.

10 Señor Oreja addressed the Cortes on the subject of the Lisbon Agreement on 16 April 1980.

11 Not without reason was Andalucia, one of the most economically backward of the regions of Spain, known as 'Spain's Third World'.

12 At this time the law precluded all non-British nationals from playing an active part in trade union affairs, from founding their own trade unions and placed restrictions upon the right of abode, purchase of property and ownership of businesses.

13 Although excluded from such contributions to Brussels, Gibraltar did not have representation in the European Parliament.

14 A point made by Mr Eldon Griffiths in the House of Commons on 14 April 1980.

15 At this time the provisional date for entry to the EC was set for January 1984.

16 It should be remembered that Spain was one of three nations seeking membership of the EC; the others being Portugal and Greece. French objections focused on Spain and Portugal and were not directed at Greece.

Addressing French farmers' leaders in Paris on 5 June 1980, President Giscard d'Estaing argued that recent deliberations within the Community had 'demonstrated clearly that the integration of certain new members is still not complete'. He asserted that, 'it does not seem possible to me to compound the problems and uncertainties related to the prolongation of the first enlargement with those created by new accessions. That is why, bearing in mind the attitude of certain of our partners since the beginning of this year, it is necessary that the Community should give priority to completing the first enlargement before it can be in a position to undertake a second'.

Although the President did not say so explicitly it was assumed that he was referring to the dispute over Britain's contributions to the Community budget and to the fact that the May 1980 settlement of that dispute had been for the two years 1980 and 1981, and that the Community had pledged itself to undertake a radical reform of its budgetary system with a view to introducing revised arrangements by 1982.

The President was also thought to be voicing the long-standing anxieties in French political quarters that French farming interests could be adversely affected by Portuguese and Spanish participation in the Community's common agricultural policy.

There was also speculation that President Giscard d'Estaing's opposition to the timetable for Portuguese and Spanish accession to the EC was not unconnected with the French presidential elections which were scheduled for May 1981. It was especially noted that both the Gaullist Rassemblement pour la République (then a member of the governing coalition) and the Communist Party (then in opposition) were emphasising the damage which would be inflicted upon farmers in the south of France should Portugal and Spain join the EC.

17 Herr Schmidt's view was made public on 9 June 1980.

18 This ruling applied to both Portugal and Spain.

9 OTHER PRIORITIES

1 In June 1980, some eight months prior to the attempted coup, Señor Oreja had implied that Spanish membership of NATO had been removed from the foreign policy agenda because some high ranking military officers had opposed Spanish membership of the Organisation unless there was progress on Gibraltar. In particular, the officers had objected to the use of the naval facilities on the Rock by NATO forces without prior approval having been sought from, and given by, the Spanish Government as they felt that such usage was incompatible with the Spanish claim to sovereignty over Gibraltar.

2 Prime Minister Calvo Sotelo made this revelation in a meeting with representatives of the press in Madrid on 24 April 1981. He also stated that the Spanish armed forces were unanimously in favour of the Government's intention to join NATO. He ruled out the possibility of a referendum on the question of Spanish membership.

3 The five-man Select Committee delegation was led by Sir Anthony Kershaw.

4 *El Pais*, 12 July, 1981.

5 *El Alcazar*, 12 July, 1981.

6 Lord Carrington and Señor Perez Llorca met in Brussels on 13 July 1981.

7 Spaniards did not enjoy the same rights in Gibraltar as the citizens of EC countries in respect of housing and business tenancies and work permits.

8 Lord Carrington was on holiday in Spain at the time. He was accompanied to the Spanish Foreign Ministry by the British Ambassador to Spain, Mr Richard Parson. No prior announcement was made of the meeting which was described by the Foreign and Commonwealth Office in London as 'private'.

9 The 1976 Treaty was due to expire on 21 December 1981. That Treaty, which superseded the 1970 Agreement of Friendship and Cooperation, which itself had been based on previous US–Spanish Agreements dating back to 1953, allowed the US the continued use of four Spanish air and naval bases at Torrejon, Saragossa, Moron and Rota in return for US loans and grants amounting to $1,120 million.

10 The extension was approved retroactively by the US Senate on 18 November 1981 and by the Spanish Congress on 16 March 1982.

11 This concern was expressed by Denmark, the Netherlands and Norway.

12 On 12 December Tass denounced the provisional acceptance of Spain into NATO and warned that the Soviet Union might take 'appropriate steps' in retaliation and maintained that the more disturbed the balance of forces in Europe became then the greater was the prospect of a heightening of international tension.

13 *Diario-16*, 31 August 1981.

14 There were 34 amendments tabled at the Committee stage of which 16 were tabled by the PSOE. All 16 were rejected by the Committee but an amendment proposed by the Catalan representatives that the Government should not accept commitments involving the stockpiling

or installation of nuclear weapons on Spanish soil was carried. An amendment moved by The Basque Nationalists (PNU) and supported by the Socialists and Communists which called for Spain to adhere within a year to the 1968 Nuclear Non-Proliferation Treaty, which Spain had neither signed nor ratified, was lost.

15 The Union of the Democratic Centre (UCD), the governing party, the Democratic Coalition (CD), the Basque Nationalists (PNV) and the Catalan minority gave their support to the resolution. Opposing it were the Socialist Workers' Party (PSOE), the Andalucian Socialists (PSA) and six Independents.

16 The Treaty of Washington was the treaty under which NATO had been established in 1949.

17 Voting for, were the UCD and PNV. Voting against, were the PSOE and the six independents.

18 The political parties opposing Spanish membership of NATO had fought hard. According to the Civil Governor's Office in Madrid, some 100,000 people had supported a rally in the city sponsored by the PSOE, PCE and trade union and civil organisations on 15 November 1981; the organisers claimed that 500,000 had taken part. Rallies had been held earlier; in Madrid on 5 July and 4 October, and in Barcelona on 6 December. By 10 December, the PSOE had collected and submitted 600,000 signatures to the Government calling for a referendum on Spanish membership of NATO.

19 In London it was conceded that there might be some criticism of this proposal from allied navies which would have preferred the retention of British naval facilities in case they were needed in any future emergency. However, there could be little doubt but that Gibraltar no longer possessed its former strategic importance for Britain as it withdrew from its previous global commitments.

20 Sir Joshua Hassan was speaking immediately after Mr Nott's announcement was reported in Gibraltar.

21 The Ministry of Defence estimated the current annual cost of the dockyard to be £10 million.

22 Reduced operating hours at Gibraltar airfield were estimated by the Ministry of Defence to be likely to result in an annual saving of £1–1.5 million.

23 The British officials were in Gibraltar from 24–6 November 1981.

24 It was later announced that Sintra, Portugal, would be the venue for the Anglo-Spanish talks scheduled for 20 April 1982.

25 On 12 February 1982 it was announced in Madrid that Señor Salvador Camino Crespo had been appointed as Under-Governor and that the Spanish Government had allotted 13,000 million pesetas, approximately £6.5 million, for investment in the Campo during 1982–4.

26 The sovereignty of the Falkland Islands had long been the subject of contention between Argentina and Britain.

27 Señor Calvo Sotelo's observations were made on 4 April 1982; the day after the Argentinian invasion of the Falkland Islands.

28 *El Pais*, 3 April, 1982.

29 Lord Carrington resigned his office on 5 April, 1982.

30 The PSOE motion was tabled on 28 May 1982. The motion had probably been tabled by the PSOE to ensure consistency with its previous stand on Spanish accession to NATO.
31 Spanish accession to NATO followed the ratification by the existing member countries of the Spanish Protocol of Accession which had been signed in December 1981. There were to be two belated demonstrations opposing Spanish membership on 6 June 1982. The first, a well attended one, was a march from Madrid to the US air base at Torrejon de Ardoz; the second, much smaller, was staged in Barcelona.
32 Mr Francis Pym told the House of Commons on 22 June that it would be 'inconceivable' for Spain to join the EC (a date unofficially set for January 1984) while the frontier between Spain and Gibraltar remained closed. He assured the House that there would be no change in the status of Gibraltar without the full agreement of its people, and expressed his confidence that the frontier would eventually be reopened.
33 With an unemployment rate of 25 per cent, La Linea, the Spanish mainland town closest to Gibraltar, had the highest unemployment rate in the whole of Spain.
34 Señor Felipe Gonzalez assumed the office of Prime Minister on 2 December 1982.
35 On 11 December, access was restricted to British passport holders resident in Gibraltar and to Spanish citizens. On 10 January 1983 the Spanish Ministry of the Interior extended access to include about two hundred Gibraltarians holding British passports but who had been resident in Spain since the frontier had been closed in 1969. These restrictions also seem to have been intended to protect the trading position of the Spanish enclaves of Ceuta and Melilla, in north Africa, and Malaga airport; it was feared that the latter might experience a decline in traffic if British and other foreign tourists flew to Gibraltar and then crossed the border into Spain on their way to the holiday resorts of the Costa del Sol.

10 HOPE RENEWED

1 No precise date for the next round of Anglo-Spanish talks had been set at this time.
2 See the letter of Gibraltar's Chief Minister to *The Times*, 1 February 1983.
3 These observations were made by Lieutenant-General James Thompson, Chief of Staff at the Naples Headquarters of the Allied Forces Southern Command.
4 Mr Pym's statement to the House of Commons was made on 10 March 1983.
5 Señor Moran was in London on 16–17 March 1983.
6 Eleven warships and two submarines docked at Gibraltar on 13 April 1983. The Spanish foreign Ministry made no comment when, later in the year, on 12–15 September, seven Royal Navy ships taking part in the Orient Express exercise put into Gibraltar; HMS *Invincible* was part of that squadron.

7 Sir Richard Parson was summoned to the Spanish Foreign Ministry on 9 April and again on 12 April 1983.
8 Trade union proposals for the naval dockyard were more wide ranging.
9 Señor Moran addressed a press conference at The Hague on 14 July 1983.
10 Sir Joshua Hassan returned to London on 26 July 1983. A week earlier, Mr Ian Stewart, Under Secretary of State for Defence Procurement, had visited Gibraltar for talks with representatives of the Gibraltar Government.
11 The new date for the closure was now set at 31 December 1984.
12 All of these assurances were conditional upon the adoption of new working practices being introduced into the dockyard.
13 Señor Moran was addressing a summer course on contemporary Spanish literature at San Roque, intended to bring the citizens of Gibraltar and Spain closer together.
14 The shipyards at Cadiz, Huelva and Lisbon were cited as regional rivals to Gibraltar. All received large subsidies from their respective Governments.
15 There was opposition to this Spanish request by, among others, West Germany. The free movement of labour was a question which had significance for Gibraltar where it was feared that the Rock's total of 10,000 jobs would become the object of fierce competition between the inhabitants and workers from Spain. This was a matter of great concern to the trade unions in Gibraltar which felt an obligation to ensure that priority was not only given to the local inhabitants but also to the Moroccan workers who had helped the Rock through the difficult days when it had experienced a severe labour shortage in the wake of the Spanish closure of the frontier in 1969.
16 King Juan Carlos also linked Gibraltar to the Anglo-Chinese agreement on Hong Kong when he made a speech welcoming the President of China, Mr Li Xiannian, on the latter's official visit to Spain.
17 In accordance with EC regulations, full freedom of movement of workers in the Community would not become applicable to Spain until seven years after final accession.

11 PARTIAL RESOLUTION

1 The Brussels Agreement had set the date for the reopening of the frontier as 'not later than 15 February 1985'.
2 Señor Borbolla emphasised that priority would be given to developing the infra-structure of the Campo region.
3 One of the questions on which clarification was sought was that of tariffs. Spain was expected to set these at the same level as at its other frontiers with France and Portugal but a decision was needed as to whether or not Gibraltar should be exempt from the 1954 New York Convention which required a minimum 24-hour absence from a country before travellers could bring a limited quantity of duty free goods with them on their return.
4 Spain's entry to the EC was scheduled for 1 January 1986.

5   All eight Government members voted for the Bill; all seven Opposition members voted against.

6   The street protest and delivery of the letter took place on 31 January 1985.

7   Sir Joshua Hassan met with Sir Geoffrey Howe, in London, on 30 January 1985.

8   Señor Moran's observation was made on 6 February 1985; the day of his return from Geneva.

9   By early June 1985 it was estimated that 36,000 visitors per week were crossing into Gibraltar from the Spanish mainland, that 650 new jobs had been created on the Rock and that the number of registered unemployed stood at 370.

10 The airport at Gibraltar, used by both civil and military aircraft, was constructed during the Second World War on the isthmus which, according to the Spanish Government, was land not ceded under the Treaty of Utrecht but which had been illegally annexed by Britain in 1908-9.

11 In doing so, Sir Geoffrey Howe became the first British Foreign Secretary to visit Gibraltar since 1971.

12 The Lower House of the Cortes ratified the EC accession treaty on 26 June 1985.

13 The proposals were those mentioned by the then Spanish Foreign Minister, Señor Moran, during the Geneva meeting of 5 February 1985.

14 Spain formally entered the EC on 1 January 1986. A referendum in Spain on 12 March 1986 on whether the country should remain in, or withdraw from, the Organisation produced a vote in favour of continued membership.

15 The British passed an *aide mémoire* to the Spanish Government containing details of this alleged incursion on 2 April 1986.

16 King Juan Carlos made the first state visit to Britain by a reigning Spanish monarch for 81 years. The state banquet was held at Windsor Castle on 22 April 1986.

17 King Juan Carlos addressed the two Houses of Parliament on 23 April 1986.

18 King Juan Carlos addressed the UN General Assembly on 22 September 1986. Speaking of Gibraltar he said that 'Spain maintains, vigorously and with the weight of the reason inherent in its cause, the will to find a rapid solution to the problem of Gibraltar, so that the Rock can be reintegrated into the Spanish national territory. A new chapter had opened in since the Brussels declaration . . . and since the Governments of the UK and of Spain decided . . . in Geneva to resolve the problem in all its aspects, including that of sovereignty through negotiation'.

19 The four-man British team was led by an official at the Department of Transport, Mr David Moss. The talks were held on 10-12 February 1986.

20 The British ceremonial military guard had been posted at the frontier between Gibraltar and Spain since the 1730s. Despite Gibraltarian protests the guard was not re-introduced.

21 These figures were made public on 5 October 1986.

22 The Spanish pensioners' original contributions and interest was stated to amount to £4.5 million.

23 Admiral Pellini's visit to Gibraltar had already been postponed following a protest by the Spanish Government to a visit made to Gibraltar, on 8 October 1986, by the Supreme Allied Commander Europe, General Bernard Rogers.

24 Sir Joshua Hassan was present at the talks in London between the British and Spanish foreign ministers and he also had a meeting with the British Prime Minister on 14 January 1987.

25 Señor Ordonez addressed the press at the end of the talks in London on 14 January 1987.

26 Mr Eggar mentioned tourism, the environment, culture and sport, public health and education as areas in which contacts between Gibraltar and Spain had developed satisfactorily since the Brussels Agreement and the reopening of the frontier by the Spanish authorities.

27 The number of banks in Gibraltar increased from 4 in 1980 to 15 by 1987.

28 A. and P. Appledore had been granted a 10-year management contract from the date of Gibraltar Shiprepair's creation on 31 December 1984.

29 The Gibraltar Government had invested £2 million to secure the commercial future of the yard during the first eight months of 1987 but had been prevented from giving further financial aid by an EC directive on shipyards.

30 EC Transport Ministers met in Luxemburg on 24–5 June 1987, to try to produce an agreement on the deregulation of air transport.

31 The two Gibraltarian leaders were invited to Strasburg by Lord Bethell, Conservative Euro MP for London North West and Chairman of the Gibraltar in Europe Representation Group. The visit had been postponed on several previous occasions as a result of Spanish objections and eventually took place only after it had been downgraded in status from an 'official' to an 'unofficial' visit.

32 These resolutions were approved on 17 November 1987.

33 These were all issues covered in the Brussels Agreement and upon which few real signs of progress had been evidenced.

34 The agreement on the Algeciras–Gibraltar ferry service and on land transport was contained in a separate joint decision.

35 The British Government had opposed joint administration of the airport because this would have enhanced the claim of the Spanish Government to sovereignty over Gibraltar.

36 Sir Joshua Hassan resigned on 11 December 1987. He had been the dominant personality in Gibraltarian politics for over 40 years and had led the AACR–GLP since 1947. He had served as Gibraltar's first mayor in 1945 and had been its first Chief Minister in 1964. He had lost only one election, in 1969, when the now defunct IWBP had enjoyed a period in office which ended in 1972. Throughout his long political career Sir Joshua had been an ardent opponent of the Spanish claim to sovereignty over Gibraltar.

INDEX